The HMS Victory Story

The HMS Victory Story

John Christopher

The
History
Press

Published in the United Kingdom in 2010 by
The History Press
The Mill · Brimscombe Port · Stroud · Gloucestershire · GL5 2QG

British Library Cataloguing in Publication Data
A catalogue record for this book is available from the British
Library.

Hardback ISBN 978-0-7524-5605-8

*Half title page: 'Death of
Nelson at the Battle of
Trafalgar' by Benjamin
West. (National Archives
of Canada)*

*Title page: 'The Battle
of Trafalgar' by William
Clarkson Stanfield. (US
Library of Congress)*

➤ *This stern
ornamentation replaced
the original galleries in
Victory's 1800 refit.*

Typesetting and origination by The History Press
Printed in Italy

CONTENTS

ACKNOWLEDGEMENTS

Most of the images in this book have either come from the archives of Airship Initiatives Ltd, or are new photographs taken by the author. Others are from the US Library of Congress, the National Archives of Canada, Paul Mckeown, Campbell McCutcheon (CMcC) and StarsBlazkova.

I am very grateful to the staff at Portsmouth's Historic Dockyard for their kind assistance. I would also like to thank my wife, Ute, for her continued support and many hours of proof reading, and my children Anna and Jay.

A number of sources have been consulted in the production of this book, including:

Sea Life in Nelson's Time by John Masefield, *Men of Honour – Trafalgar and the Making of the English Hero* by Adam Nicolson, *The Habit of Victory* by Captain Peter Hore, *Nelson* by Roy Hattersley, *The Nelson Companion* edited by Colin White, *The Book of British Ships* by Frank H. Mason, and *Shipping Wonders of the World* edited by Clarence Winchester.

Heart of oak are our ships, jolly tars are our men,
We are ready; Steady, boys, steady!
We'll fight and we'll conquer again and again.

Heart of Oak, with words by David Garrick, became the official march of the Royal Navy

HEARTS OF OAK

Towards the end of the eighteenth century the old world order was undergoing something of an upheaval. On the other side of the Atlantic a bunch of colonists had the temerity to declare independence, while in Europe a bloody revolution and the rise of Napoleon Bonaparte was stirring up a storm. The result was a succession of conflicts and, consequently, Britain's Royal Navy was far bigger than it would otherwise have been during peacetime. This meant that there were more than enough glorious battles to be fought and great victories to be won to meet the needs of any aspiring hero.

'Cometh the time, cometh the man.' This adage could not have been more apt when applied to Britain's most famous sailor, Admiral Lord Nelson. In the hero-making

➤ Bowsprit and union flag, with the figurehead removed for restoration.

➤➤ Admiral Lord Nelson 1758–1805.

➤➤➤ Long past her natural shelf life, Victory moored in Portsmouth, c.1900. (US Library of Congress)

business here is a prime example of the right man being in the right place at the right time. If his times had been any less eventful then young Horatio might have stayed in Norfolk and become a clergyman like his father.

Not only did he look better in a uniform than a dog collar, Nelson had extraordinary charisma and leadership qualities. The women adored him, his men would follow him into hell and back. What's more, he had the foresight to die young – he was only forty-seven years old at Trafalgar – and to die gloriously, two essential elements in the creation of such a potent hero myth.

Equally as famous is his greatest flagship, HMS *Victory*. It was not the only ship Nelson served on, and he was not the only admiral to hoist his flag on *Victory*, not by a long shot, but it was his ship at the most famous moment in his career and in the history of naval conflict. It also happens to be an extraordinary survivor, the lone one of her era. Following a hammering at the

Battle of Trafalgar, *Victory* was refitted, virtually rebuilt, and went on to serve in two Baltic campaigns. Miraculously, thanks largely to the sentimentality of key figures within the Admiralty, she escaped the breakers yard. Unlike HMS *Temeraire*, which fought alongside *Victory* at Trafalgar and is famously depicted in J.M.W. Turner's oil painting 'The Fighting Temeraire tugged to her last berth to be broken up, 1838'.

Most wooden ships of that time were only expected to last a dozen years, maybe twenty. *Victory* has been around for over 200 years and since the 1920s has attracted thousands of visitors to the historic naval dockyard at Portsmouth. She is an incredible ship; a unique time capsule of life in Nelson's navy. When I saw her for the first time I was impressed by her size – much bigger than I had expected – a

wall of black and yellow-ochre peppered with gun ports. She must have been an incredible sight when she was in full sail, and the concept of thirty or forty of these fighting ships slogging it out at close range is almost beyond imagination.

Did you know?
Horatio Nelson suffered terribly from seasickness throughout his naval career.

Above all else, HMS *Victory* is an icon of those values which were so important in Nelson's time – patriotism, heroism and an unwavering sense of duty. This is why Winston Churchill was photographed visiting the ship during the dark days of the Second World War. He wanted to assure people she was undamaged by the air raids. He also recognised the powerful symbolism she evokes, as did Hitler who talked of taking Nelson's Column back to Berlin as a souvenir after the invasion of Britain. But Nelson and *Victory* had repelled one would-be invader and Churchill would thwart another.

Victory's heart is of strong English oak. When the celebrated actor David Garrick wrote the words to the song *Heart of Oak* he was not only referring to the great ships-of-the-line, but he was also alluding to the men who served on them. This is their story too.

This day will be launched His Majesty's Ship the *Victory*, estimated the largest and finest ship ever built. Several of the Lords of the Admiralty, Commissioners of the Navy, and many persons of quality and distinction, are expected to be present, for whose reception great preparations are making through the town.

London Public Advertiser, 7 May 1765

The building of a first-rate ship-of-the-line, a battleship for King George III's Royal Navy, was an enormous undertaking – the equivalent of building an aircraft carrier in modern times. Far more than just sailing ships, these wooden battleships were intricate and complicated fighting machines which could stay at sea for up to four months to slog it out against the nation's enemies – and there were a lot of those – while sustaining a huge crew of up to 850 men. Ships such as HMS *Victory* were the largest moveable man-made objects of their age, and building one demanded huge resources in terms of money, materials, manpower and time.

HMS *Victory* was conceived during the Seven Years War of 1756–1763, an international conflict of imperial rivalry that would set the balance of power well

13

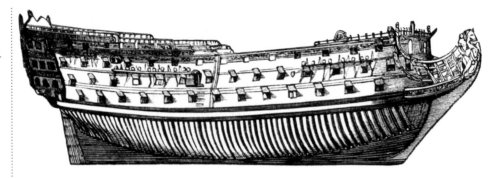

Did you know?

The expression 'Letting the cat out of the bag' refers to the cat-o-nine-tails whip used as a punishment. When it came out of its bag there was going to be trouble for someone.

into the next century. Two years into the war, in December 1758, King George II's ministers decided to build twelve ships-of-the-line, including a first-rate ship. Royal Navy ships were 'rated' according to the number of guns carried, and a first-rate ship such as *Victory* had 100 guns or more. A second-rate ship, no slight intended, carried between ninety and ninety-eight guns, a third-rate had sixty-four to eighty, and so on down to a sixth-rate with only twenty-four to twenty-eight guns.

In December 1758, the commissioner of the naval dockyard at Chatham received instructions from the Admiralty to prepare a dry dock for the construction of a new ship. By 1759 the war was going well for Britain with a number of victories – notably on land at Quebec and the naval battles at Lagos and Quiberon Bay – so the

warrant for the construction of the new ship was signed by the Navy Board on 7 July 1759.

The design of the ship, based on the plans of HMS *Royal George* launched from the Woolwich Dockyard three years

earlier, was put in the hands of the naval architect Sir Thomas Slade, the Senior Surveyor of the navy. The big first-rate ships, such as *Victory*, were few and far between. The navy preferred smaller and more manoeuvrable ships in general, and during the whole of the eighteenth century only ten 100-gun ships were built. When completed, *Victory* would be 226ft (69m) long overall, bowsprit to taffrail, with a keel length of 151ft (49m) and maximum breadth, or beam, of 51ft (15.7m). The number of guns dictated the size of the ship's company, and hence the space needed for accommodation and stores. A fully crewed first-rate ship would normally have 850 men – a term which actually included a number of boys and a few women most likely – although the crew for *Victory* usually numbered 820.

The first part of the building process, the laying down of the keel, took place at the Old Single Dock at Chatham on 23 July 1759; an auspicious moment marked by the presence of Prime Minister William Pitt the Elder, no less. The keel is the backbone of the ship from which the main timbers would spread upwards to form her frame.

The figures for the amount of wood used to build the ship are staggering. Around 2,000 trees were felled, the equivalent of 100 acres of woodland, to supply 300,000 cubic feet of timber. It

◀◀ *Engraving of an eighteenth-century carpenter using hand tools.*

◀ *Instead of the conventional method of building wooden ships on a slipway, shown here, Victory was built in dry dock at Chatham.*

A piece of one of Victory's original masts, constructed from several wooden sections, now on display on the middle gun-deck.

Eighteenth-century mast-maker.

was mostly oak, the remainder being elm and fir for the hull, or fir, pine and spruce for the masts and yards. This timber would have been in store and left to season for some fourteen years; a process that contributed greatly to the ship's strength and longevity.

Working the timber was done entirely by hand, and at first around 250 men were employed on Victory's construction. In the pre-industrial times these shipwrights came with a variety of specialist skills which had hardly changed for centuries, from the woodworkers and carpenters to, later on,

◄ *Victory has 768 wooden pulley blocks on the rigging, plus a further 628 for the guns.*

the rope and sail makers. A name for the new ship was selected from a list of seven reserved for first-rates. *Victory* was the only one not in use, and despite some reluctance within the Admiralty Board as the previous *Victory* had been lost with all hands in

Making rope by hand.

Did you know?

Victory was considered one of the fastest ships of her time, but even so she could only manage a maximum of around 8–9 knots. That's the equivalent of 10mph (16.1km/h).

1744, the name was finally agreed for the new ship in October 1760. However, within twelve months the workforce had been reduced as the changing fortunes of the war with France made the need to complete her less urgent.

Once the frame of a wooden ship was completed the normal practice was to cover it up and leave it to 'season in frame' for up to twelve months, but with the Seven Years War coming to a long drawn-out conclusion, *Victory* was left for almost

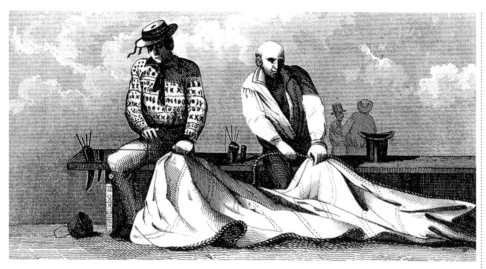

three years. Work finally resumed in the autumn of 1763, and she was towed out of the dock on 7 May 1764 to be ballasted with 34 tons of shingle.

With no role for her *Victory* was placed in 'ordinary', in other words kept in reserve, dismasted, covered over and moored in the River Medway until needed. It was a full twelve years later before the Admiralty ordered the ship to be completed for service as Britain's old enemy, France, had joined in the American War of Independence and

LAUNCHED AT CHATHAM DOCKYARD 7TH. MAY 1765.

EXTREME LENGTH 226-6′.   LENGTH OF KEEL 151′-3″.
EXTREME BEAM 52-6″.   DEPTH OF HOLD 21-6″.
LENGTH OF GUN DECK 186′.   TONNAGE 2162 TONS, DISPLACEMENT TONNAGE 3500TONS

ARMAMENT.

LOWER DECK 30 LONG 32 PDR. MIDDLE DECK 30 LONG 24 PDR. MAIN DECK 32 LONG 12 PDR. UPPER DECK 12 SHORT 12 PDR.

▲ *This decorative panel on the middle gun-deck celebrates Victory's launch on 7 May 1765.*

further conflict was looking inevitable. In Europe, meanwhile, great changes were afoot and the French Revolution was about to upset the political apple cart, bringing new heroes and villains to the fore.

On 13 April 1778, *Victory* finally left Chatham and made sail as Admiral Augustus Keppel's flagship, the start of an illustrious career that would make her the most famous ship in the world.

Did you know?

There was no toilet paper in Nelson's time. A tow rag was a piece of frayed rope which a sailor would clean themselves with after visiting the head. It was then trailed in the water ready for the next man.

Something must be let to chance; nothing is sure in a sea fight above all.

Nelson, before the Battle of Trafalgar

HMS *Victory* is a ship of contradictions. From the outside she is surprisingly large, while inside the decks get progressively cramped. Visitors enter the ship on the port side and step straight into the middle gun-deck before being led on a circuitous route up, down and around through the other decks. For the sake of simplicity this guide begins at the top, at the stern, and works downwards:

POOP DECK

This short deck is raised up at the stern of the ship and, contrary to suggestions of countless schoolboys, it takes its name from the Latin *pupis* meaning after deck or rear. Its main purpose was as a viewpoint and a signalling platform and it is here that the signal flags are stored in signal lockers. The ropes controlling the yards or spars and sails of the main mast and

23

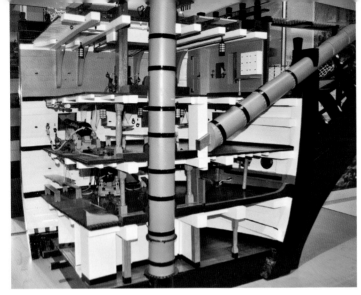

cabin, although sliding shutters maintained privacy when required.

▲ *Cutaway model of Victory's bow section showing from the top; upper gun-deck in ochre, the middle gun-deck, lower gun-deck and orlop. Note how the foremast passes down through the decks.*

mizzenmast were operated from the poop. At the rear are three signal lanterns which burned whale oil to enable ships of the fleet to see each other's position to either stay in formation or avoid collisions. The skylight in the middle of the poop provided additional light for the captain's dining

QUARTERDECK

Just ahead and below of the poop, the quarterdeck has been described as the nerve centre of the ship. It is from here that a captain controlled his ship, and this is where Nelson directed the fleet at Trafalgar. Unfortunately, this made the quarterdeck

Quarterdeck, looking aft.

◄◄ *Entrance on to the middle gun-deck, on the port side.*

◄ *Quarter deck looking aft towards the ship's wheel and above it the poop deck.*

a prime target for enemy marksmen and a brass plaque indicates the spot where Nelson fell. The quarterdeck has twelve short 12-pounder guns.

At the back of the quarterdeck, sheltered under the edge of the poop, are the ship's wheel and the compass binnacle. As many as four men where needed to operate the

wheel. In bad weather it could take up to eight. Beyond the wheel is the entrance to the captain's cabin. A little smaller than the admiral's quarters, it is divided into three areas with the day cabin, dining cabin and a bed space.

◄ The upper gun-deck is surprisingly spacious and was used as a workspace.

FOC'SLE

Located at the front of ship, the sails and yards on the main and fore masts were controlled from the foc'sle, or forecastle, and the anchors were operated from here. It was also an area the crew could use when

▶ *Situated behind the ship's wheel, the captain's dining room was not quite as well appointed as the admiral's quarters.*

off duty. The foc'sle has four guns, two medium 12-pounders and two 68-pounder carronades which were immensely powerful short-range guns. (A ship's guns are never referred to as cannon.) The ship's bell is on the foc'sle and this was essential in maintaining time on the ship, especially at the beginning and end of each watch. To ensure accuracy, a pair of sandglasses were used at the start of the watch.

UPPER GUN-DECK

Down one level you come to the first of the gun-decks, and it is surprisingly spacious. At the rear is the great cabin which is divided into three areas, just as with the captain's cabin – day, dining and bed spaces. Before battle the screens which divided it from the rest of the deck were removed to turn it into part of the gun-deck. The main area of the gun-deck had thirty 12-pounder guns facing out through the gun ports. The central space was kept open to provide a well-lit workspace for the crew.

The sick berth is at the front of this deck, as far away from the living areas as possible to reduce the spread of diseases, partitioned by canvas screens. This too could be turned into gun space in battle, and the injured were treated down below in the orlop.

MIDDLE GUN-DECK

Another gun-deck, with twenty-eight 24-pounders. At the stern are a number

▲ *External view of the admiral's quarters.*

the public. Towards the bow is the galley, made up of two sections – the pantry and a cooking area equipped with an iron cooker known as a Brodie stove. The ship's capstan heads are also located on the middle deck. One at the stern was used to raise the anchors, while the 'jeer' capstan in the centre of the deck was used to lift stores as well as raising masts and yards.

LOWER GUN-DECK

Down another level and this gun-deck houses thirty 32-pounders. It served as the living area for the majority of the crew and about 460 men would sleep on this deck with only the regulation 14in (35cm)-wide space for their hammocks. During the day the hammocks were stowed away and at meal times the men ate on tables slung from the ceiling, or on folding tables set

The admiral's great cabin, at the rear of the upper gun-deck. They would clear the furniture away when going into battle.

of small cabins for the wardroom, which served as living quarters for the naval and marine officers. This is one part of the ship still in regular use and is not open to

up in the central area. (See 'Life on board *Victory*' for more on the food and eating arrangements.)

THE ORLOP

As it is located beneath the waterline, the orlop could not be used for guns. This made it an ideal space for storage and cabins. The surgeon's cabin was located here and the purser also had a cabin right next to the stores. On the other side of his cabin was the purser's steward's room, and it was from here that the daily rations of food and drink would have been measured out and issued to the mess cooks. The steward would sleep here to prevent pilfering by the crew. The ship's two hanging magazines and the gunpowder cartridge store are

◄▲ *24-pounders on the middle gun-deck.*

▲ *The officers' wardroom, at the stern of the middle gun-deck, also doubled up as the surgeon's operating theatre.*

accessed from the orlop. They were positioned below the waterline to protect them from enemy fire.

THE HOLD

The last level, the hold, is the largest storage area on the ship with space for up to six months' worth of provisions. Most items were stored in water barrels called 'leaguers', each one holding 150 gallons (682lts) of water. It was in one of these barrels that Nelson's preserved body was brought back to Britain. At the bottom of the ship is the ballast used to counteract the weight of the guns and the tall masts. It consisted of 200 tons of pig iron and shingle, a material that could be moved around easily to adjust the ship's trim. The shot lockers, where the shot for the guns was stored, are also in the hold because of their weight.

MASTS, YARDS AND RIGGING

Victory has three vertical masts plus the bowsprit which projects out from the bow. From front to back they are the foremast, the main mast and the mizzenmast. Each one is made up of three overlapping sections. The lowest is called the mast, the middle the top mast, and at the top there is the topgallant mast. The masts

are made of a flexible wood such as fir, pine or spruce. The sails were hung from the cross pieces, the yards, and each mast had three yards plus another yard that could be raised if needed. Where the masts meet the top masts there are platforms called tops. These help to spread the standing

The cable tier in the orlop where the long ropes for the anchors were stored.

33

▲ Carpenter's workshop in the forward part of the orlop deck.

➤ Pig iron and shingle ballast at the bottom of the hold.

➤➤ Each mast had a platform, a top, which helped to spread the standing rigging and provided a well-protected position for the marines with their muskets.

rigging holding the masts upright and were used as platforms to fire upon the enemy in battle.

Victory has approximately 26 miles (42km) of rigging. Standing rigging supports the masts and doesn't move, while the running rigging was in constant use, hoisting the yards and sails. There are approximately 768 wooden pulley blocks on the running rigging.

If any ship or vessel be taken as prize, none of the officers, mariners, or other persons on board her, shall be stripped of their clothes, or in any sort pillaged, beaten or evil-intreated, upon the pain that the person or persons so offending, shall be liable to such punishment as a court martial shall think fit to inflict.

Articles of War on board a Royal Navy ship, 1757

HMS *Victory* did in fact have a noteworthy service life before the Battle of Trafalgar and, for that matter, before the arrival of Horatio Nelson. By the time of her completion it was obvious that war with the French was inevitable. She was sent from Chatham to Portsmouth to hoist the flag of Admiral Augustus Keppel, Commander-in-Chief of the Channel fleet.

Keppel soon had his new flagship in action, at the first Battle of Ushant – named after an island on the north-westernmost point of France, between the mouth of the Channel and the Bay of Biscay, known to the French as Ouessant. With a force of thirty ships, Keppel put to sea on 9 July 1778 and they sighted the French fleet of twenty-nine ships 100 miles (160km) to the west of Ushant. In

 Victory's *three masts, looking forward from near the stern on the starboard side; missen mast, main mast and fore mast.*

command of the French was Admiral Louis Guillot, who had been ordered to avoid battle, but was cut off by the British from the shelter of the port of Brest. (Two of Guillot's ships made a successful dash for port leaving him with twenty-seven.) In

▲ Admiral Augustus Keppel took Victory into action at the first Battle of Ushant. (US Library of Congress)

➤ 'An English-jack-tar giving monsieur a drubbing.' Contemporary cartoon celebrating the first Battle of Ushant. The inn sign shows Admiral Keppel and the Victory is in the background. (US Library of Congress)

shifting winds and a heavy rain squall, the opposing fleets manoeuvred for advantage.

The British, for their part, were more or less in column, while the French fleet was in some confusion until they managed to pass along the British line. Shortly before midday Victory opened fire on the 100-gun Bretagne, closely followed by the 90-gun Ville de Paris. The leading British ships escaped with little loss, but the rear division, commanded by Sir Hugh Palliser, suffered more. When Keppel signalled to follow the French ships Palliser did not conform and the action ceased. On their return to England Keppel resigned, faced a court martial but was cleared of dereliction of duty, and Palliser was severely criticised by the official enquiry.

In 1780, Victory's lower hull was sheathed with copper, a common if expensive practice to protect it from the teredo shipworm, and this helped to prevent the

formation of barnacles which accumulate in a crust many inches thick and reduce a ship's speed by several knots. *Victory* was considered to be one of the finest first-

➤ Victory's waspish yellow-ochre and black colour scheme was implemented after the 1800 refit.

➤➤ At the Battle of St Vincent Admiral John Jervis had Victory as his flagship, while Commodore Horatio Nelson was in command of HMS Captain.

➤➤➤ A relief depicting Nelson receiving the surrender of the Spanish ship San Nicholas at the Battle of St Vincent in 1797.

41

rates in the navy and possessed remarkable sailing qualities for a ship of that size.

A second Battle of Ushant saw *Victory* in action again, this time as the flagship of Rear Admiral Richard Kempenfelt and under the command of Captain Henry Cromwell. On 12 December 1781, a British squadron of twelve ships-of-the-line, plus one fourth-rate ship and five frigates, was closing in on a French convoy out of Brest on its way to America. Kempenfelt suddenly realised that the convoy was protected by twenty-one ships-of-the-line under the command of Comte de Guichen. This escort was downwind of the convoy, and the British ships swept down and captured fifteen of the convoy's ships, although Kempenfelt felt his force was inadequate to tackle de Guichen's warships. The French convoy was dispersed in a gale shortly afterwards.

The following year, 1782, *Victory* led the fleet which raised the Siege of Gibraltar under Admiral Earl Howe. Over the next few years a bewildering succession of admirals followed, and by February 1797 *Victory* was the flagship for Admiral Sir John Jervis, sailing with a force of fifteen ships-of-the-line to Cape Vincent, off the southern Portugese coast. It was during the ensuing battle against the Spanish that a certain Commodore Horatio Nelson, in command of HMS *Captain*, seized the opportunity to take centre stage. (See 'Horatio Nelson – Hero of the Nile' for more on the Battle of St Vincent.)

By the end of 1797, however, *Victory* had been relegated to the undignified job of prison-hospital ship on the Medway, moored within sight of the dockyard where she had been built. Having been launched thirty-two years previously, *Victory* was showing her age and would no doubt have been retired if it was not for another twist of fate. On 8 October 1799, HMS *Impregnable* failed to live up to her name when she was lost off Chichester, having run aground on her way to Portsmouth. Unable to be re-floated she was stripped of her gear and dismantled, which left the Admiralty short of a first-rate ship. It was decided to refit *Victory*. Work started in 1800, but it was soon found that the repairs were so extensive that the ship was virtually reconstructed.

The original ornate stern, covered with galleries, was replaced by a much simpler stern with glass windows. The elaborate figurehead was replaced with the simpler crowned shield and cupids she still wears. More significantly, extra gun ports were

◄◄ Jervis congratulates Nelson after the Battle of St Vincent.

Did you know?
The masts currently on *Victory* were taken from a ship called *Shah* in the 1880s. Unlike the wooden originals these are hollow wrought-iron masts which are easier to maintain.

added, increasing her guns from 100 to 104, and her magazine was lined with copper. To complete the makeover the old red paint scheme was changed to the familiar waspish yellow and black stripes we see today. The gun ports were yellow at first, but later changed to black to create the pattern known as 'Nelson chequer', which was generally adopted by Royal Navy ships after Trafalgar. All this work cost the considerable sum of £70,933, and by the time *Victory* was ready for sea again she was as good as new.

Firstly you must always implicitly obey orders, without attempting to form any opinion of your own regarding their propriety. Secondly, you must consider every man your enemy who speaks ill of your king...

Nelson's advice to a midshipman aboard the
Agamemnon, **1793**

The story of HMS *Victory* is inexorably linked with that of Britain's most celebrated seaman, Admiral Lord Nelson. Courageous, inspirational, charismatic, patriotic, romantic – he was all of these. Vain, insecure, egotistical – he was those too. But above all else he was the consummate hero of his time.

Horatio Nelson was born on 29 September 1758 – the same year as *Victory* was ordered – at his father's rectory in Burnham Thorpe, Norfolk. The Reverend Edmund Nelson and his wife Catherine had eleven children, of which Horatio was the sixth. Theirs was a modest background with absolutely no connection with the sea apart, that is, from Horatio's maternal uncle, Captain Maurice Suckling. In all likelihood Horatio would have followed in his father's footsteps if it had not been for Suckling, and at the age of twelve he joined HMS *Raisonable* under his uncle's command

and was appointed as a midshipman in preparation for officer training.

Over the next few years Nelson found his sea legs serving on merchant ships in the East Indies, crossed the Atlantic twice, joined an expedition to the Arctic, fought a polar bear, and discovered that he suffered from chronic seasickness. After his return to England in 1776 he was made acting lieutenant on HMS *Worcester*, and the next year was made lieutenant of HMS *Lowestoft*. With the outbreak of the American War of Independence, he returned to the Caribbean and took command of the captured tender *Little Lucy*. Following France's entry into the war he was appointed as commander of HMS *Badger* at the age of twenty.

The young officer was beginning to attract attention, and in 1779 he became captain of HMS *Hitchinbroke* and was dispatched to Nicaragua as the senior officer in command of a joint force to attack the Spanish fortress at San Juan.

Nelson leaves home to go to Sea
for the first time. 1771.

The assault on the fort was a success and Nelson was praised for his initiative and leadership. Unfortunately, he was struck down by malaria and sent back to England to recover, after which he took command of HMS *Albermarle*. He spent the rest of the war in the West Indies hunting French and Spanish ships. With the coming of peace he commanded the HMS *Boreas* with the task of suppressing illicit trade with the former colonists. It was in the West Indies that he married Frances Nesbit and the couple returned to Norfolk where they had to get by on half pay, as there were no longer enough postings in the peacetime navy.

With the outbreak of the French Revolutionary Wars, Nelson was put in charge of the 64-gun HMS *Agamemnon* and sailed to the Mediterranean as part of Lord Hood's fleet. It was during his time

◀ *'Nelson leaves home to go to sea for the first time, 1771.' With his mother dead by this time, Horatio bids farewell to his grandmother.*

➤ *Battleships entering the Bay of Naples.*

in Naples that he met Emma Hamilton, the new wife of the British Ambassador William Hamilton. In July 1794, during the successful siege against French forces at Calvi, he was struck in the face by debris and eventually lost the sight in his right eye. Undaunted, he distinguished himself in actions against the French fleet the following year, earning him a promotion to commodore and a transfer to the larger HMS *Captain*.

Nelson's fame was spreading rapidly. In February 1797 he was at the Battle of St Vincent with the *Captain* as part of a British fleet of fifteen ships headed by Admiral John Jervis in the flagship HMS *Victory*. Although they were outnumbered by the Spanish, Jervis ordered his ships to sail through the two Spanish divisions to split them up. Nelson noticed that some of the Spanish ships might escape and, on his own initiative, he took the *Captain* towards them. After exchanging broadsides, Nelson

◀ *Known as a great beauty, Lady Emma Hamilton had a penchant for putting on performances of her 'attitudes', representations of classical figures.*

John Bull, the Leviathan of the Ocean; or, the French Fleet sailing into the Mouth of the Nile!

London Pub.d by W.ᵐ Holland N.º 50 Oxford S.t Decbr 12. 1798

50

in one go! The truth might not have been quite so black and white, but the glowing newspaper reports of Nelson's heroics won him the hearts of an admiring public back in England.

Then, five months later, in July 1797, Nelson led a boarding party in small boats from HMS *Theseus* to capture the Spanish treasure ship *Principe de Asturias* during the Battle of Santa Cruz de Tenerife. He was hit in the right arm by a musketball and was rowed back to *Theseus*, but the ship's surgeon was unable to save his shattered arm. He wrote afterwards:

found himself alongside the 80-gun *San Nicholas* and, according to legend, leapt across crying, 'Westminster Abbey or glorious victory!' The *San Josef* attempted to come to the aid of its compatriot, but became entangled with its rigging, upon which Nelson seized her too. Two ships

A left-handed admiral will never again be considered as useful, therefore the sooner I get to a very humble cottage the better and make room for a better man to serve the state.

➤ In 1806 Nelson's right arm is shattered by a musket ball at the Battle of Santa Cruz.

Despite his despondency he was greeted as a hero when he returned to England to recuperate. Within the year he rejoined Sir John Jervis's fleet, off the Portuguese coast, on board HMS *Vanguard* from which he commanded a detached squadron of ships to seek out and destroy Napoleon's fleet. But first he had to find it.

The French were expected to sail from the Mediterranean port of Toulon, but on the way to intercept them the *Vanguard* was damaged in severe gales. By the time the ship had been repaired and Nelson got to Toulon the French ships had already gone. The search continued along the Italian coast, and when Nelson heard that Napoleon's forces had taken Malta he was convinced that they were heading for Egypt. Accordingly the British sailed eastwards and, during the night, they

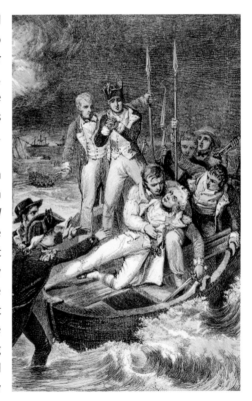

actually passed the French unaware. Nelson then ordered his ships to Sicily to take provisions on board, before returning to the eastern Mediterranean where, on 1 August 1798, he found the French fleet anchored at Aboukir Bay. Despite being outnumbered Nelson immediately launched his attack, taking the enemy off guard, destroying all but four of their ships.

The Battle of the Nile, as this action became known, had thwarted Napoleon's ambitions in the east and had seriously reduced his Mediterranean fleet. In recognition of his bold and unorthodox approach in plunging straight in to the attack, Nelson was publicly proclaimed as the 'Hero of the Nile'.

Almost immediately he was appointed as second-in-command to Admiral Parker's expedition to the Baltic to defeat a coalition of forces led by the Russians. When the British began their attack against the Danish fleet at Copenhagen, on 2 April 1801, three of their warships ran aground and Admiral Parker sent the signal for Nelson's ship to withdraw. Putting his telescope to his right eye, Nelson commented, 'I have the right to be blind sometimes. I really do not see the signal.' In the ensuing battle he fought fiercely, but after three hours both sides were left heavily damaged but unwilling to capitulate. In gentlemanly fashion, Nelson despatched a letter to the Danish commander, Crown Prince Frederick, proposing a truce, and the following day he entered Copenhagen to begin formal discussions.

Nelson was then placed in charge of defending the English Channel to keep a French invasion force at bay, but the

▲ Nelson became the pin-up of his age, and portraits such as this were mass-produced for years after his death.

The often misquoted incident when Nelson put the telescope to his blind eye at the Battle of Copenhagen, 1801.

Battle of Copenhagen.

frequent patrols proved uneventful. Operating out of English ports he was able to spend more time ashore, and this set society tongues wagging as he had obviously become cold and distant with his wife Fanny. Instead, he openly spent most of his free time with Emma Hamilton, who had given birth to their daughter earlier that year. They called the child Horatia, and as if no one had made the obvious connection they claimed that she was adopted. This period of domesticity came to an abrupt end when war broke out in May 1803. Nelson was put in command of the Mediterreanian fleet, and he had *Victory* as his flagship.

Come, brave, honest Jack Tar, once more will you venture?
Press warrants they are out; I would have you to enter
Take some rich Spanish prize, as we have done before O
Yes, and be cheated of them all, as we were in the last
  war, O.

**'Jack Tar' – ballad lamenting the plight of those pressed
into service by the press gang**

Our naval glory was built up by the blood and agony of thousands of barbarously maltreated men. It cannot be too strongly insisted on that sea life, in the late eighteenth century, in our navy, was brutalising, cruel, and horrible.

John Maesfield's account of *Sea Life in Nelson's Time*, published in 1905, paints a grim picture of conditions on board a ship-of-the-line, one characterised by appalling living conditions and a regime built on cruel discipline. But who are we to judge the past? Instead we need to cast aside our modern sensibilities and understand that life for the vast majority of ordinary working people was incredibly harsh. Whether you worked in the countryside or in the towns and cities it was little short of slavery, and a person could just as easily be hanged for a petty crime on land as a sailor would be punished for their misconduct at sea.

PLAYER'S CIGARETTES.

▶ *A Player's cigarette card showing a typical sailor's uniform at the time of Trafalgar.*

▶▶ *'Portsmouth Point', Thomas Rowlandson's cartoon of debauchery in the naval town. (US Library of Congress)*

PLAYER'S CIGARETTES.

SEAMAN. 1805.

than 90 per cent of the 92,000-odd British fatalities during the Napoleonic Wars were caused by disease, or through accidents and shipwrecks. Yet, despite the inherent risks, a life on the ocean waves was still an attractive proposition for many, lured by the promise of regular meals and drink rations, as well as the possibility of a share in any enemy prizes. Consequently, the Navy had little trouble finding sufficient recruits during peacetime. However, at periods of war they required three or four times as many men.

In order to close the gap between demand and supply, the state had a legal right to 'impress' sailors from merchant ships, or from the shore, and compel them to serve. This gave rise to the notorious press gangs who roamed the streets of harbour towns, forcibly seizing men to join the ships.

Life at sea could also be downright dangerous. Not only when shot or musket fire were hurtling your way, but more

57

A Cruikshank cartoon of an eighteenth-century naval captain. (US Library of Congress)

Officially, they were only supposed to 'recruit' seafarers, but in desperation they took any able-bodied man who crossed their path. There were many cases of men taken from their families without even giving them word of what had happened. Once on board ship, and closely guarded by the marines, some would resign themselves to their fate and accepted the enlistment bounty in the hope that this would earn some favour with the officers. But many others remained resentful at their unjust treatment, and it was partly for this reason that discipline was so harsh on board naval ships.

The Articles of War, originally established in the 1650s, and amended by an Act of Parliament in 1749 and again in 1757, governed every aspect of a seaman's life, from religious observance to matters

of general conduct. For serious crimes, the ultimate punishment was death by hanging, and for lesser offences, such as drunkenness, falling asleep on watch or refusing orders, a seaman could be sentenced to a flogging. Such punishments

Another cartoon, not of the press gang but in response to the poor treatment of naval officers, 1796.

Around seventy-five marines were based on board Victory, *their main tasks to enforce discipline among the crew, provide protection and to take on the enemy in battle.*

were always administered in front of the entire crew as an example to others. For a flogging, the culprit would be stripped to the waist, tied to a wooden grating and given as many as thirty lashes with the infamous 'cat-o-nine-tails', a short wooden stick with tails of knotted chord 2ft long. Excruciatingly painful, one blow was sufficient to rip the skin, and twelve were said to leave a man's back looking like raw meat. Having taken his punishment the victim would have his wounds washed in brine by the surgeon before returning to duties with no lasting blemish on his record.

Life on board ship was highly ordered. In charge was the captain, a god-like figure with absolute authority ruling in their own way, and a harsh one could make the lives of the crew a living hell. What was needed was a firm hand, someone who knew their duty, made the men do theirs, and in doing so commanded their respect.

The next links in the chain of command were the commissioned officers. The first lieutenant was responsible for the working

Flogging was a common form of punishment and the cat-o-nine-tails is shown here hanging beside its infamous red bag.

of the ship, the preservation of discipline and the navigation of the vessel according to the captain's orders. Supporting him were between three to eight lieutenants who took varying degrees of precedence according to the dates of their commissions. Next came the master, who controlled the sailing of the ship, the trimming and setting of the sails and guided her movements in battle. He also had charge of the stowage, the distribution of the various stores and ballast which was vitally important as the way they were positioned in the hold could affect the way the ship sailed. Bigger ships carried second masters to assist him, while smaller ships might have several master's mates.

A midshipman was an apprentice officer, usually a young gentleman taken aboard by the captain to oblige their relations. Many went to sea at the age of eleven or twelve, just as Nelson had, and spent several years learning their craft before going on to become master's mates or lieutenants. A big ship, such as *Victory*, might have as many as twenty-four midshipmen who were generally looked upon as slaves to the first lieutenant. As well as taking lessons from a schoolmaster, usually the ship's chaplain, they were expected to go aloft with the men to learn how to furl or reef a sail.

The warrant officers, or 'standing' officers, all held particular skills: the surgeon; chaplain; gunner; boatswain; purser; carpenter; sailmaker and so on. The surgeon took care of the sick, treating fevers, dressing ulcers and either bandaging the wounded or amputating their shattered limbs. The purser was a very powerful figure. In essence he was the ship's grocer and had control of the food and drink, plus

Did you know?
To ascertain the speed at which a boat was travelling, a rope with a wooden block at its end and knots tied in it would be lowered over the side. Hence the nautical measure for speed is given as knots.

other provisions such as the hammocks, bedding, tobacco and even the clothing for which the men had to pay. There were no official uniforms at this time, although some captains had a stricter dress code than others. The sailors wore long trousers which could be rolled up, and short-waisted jackets or heavily knitted jerseys. Beards among naval men did not become the fashion until later in the nineteenth century, although many of Nelson's men wore their hair long or in neatly plaited pig-tails. During voyages to the South Pacific some had also taken to the art of tattooing.

In addition to the men, there were two other categories of crew. The ship's 'boys' were young lads who entered service between the age of thirteen and fifteen and trained in the basics of seamanship. They were usually allocated the dirtiest work on the ship and some would act as servants to the officers. Woe betide the boy who was assigned to a midshipman who might be no older than them, but an officer nonetheless. Those who survived the bullying and the harshness of life at sea would go on to become ordinary seamen. Contrary to regulation, a number of women also lived aboard ship. In a few celebrated cases they served alongside the crew disguised as men, but many travelled as the wives of the officers or lived on the lower decks. In some cases they even took their children with them, or gave birth on board as Mary Buek did on HMS *Ardent* during the Battle of Copenhagen. Other 'professional' ladies poured on to the ships whenever they came into port, and in general this 'entertainment' was tolerated by the officers. Talk about turning a blind eye!

Duty is the great business of a sea officer; all private considerations must give way to it, however painful it may be.

Nelson in a letter to Frances Nisbet

Each ship-of-the-line was a self-contained wooden world. The seamen worked, ate, slept on board, some of them for years on end. Their vessel might carry provisions for four months at sea, but even in port they were often confined to the ship. This was especially so at times of war, when the number of involuntary volunteers was higher and it was feared men would desert given half the chance. For many, their time at sea felt like a prison sentence.

The structure of the day was governed by a series of watches, at least for the skilled seamen, signalled by the sounding of the ship's bell. The crew was divided into two, larboard (now known as port) and starboard, and they would takes turns as the 'watch on deck', carrying out the duties needed to sail and navigate the ship. There were seven watch periods in total and each had a name, the first being the Afternoon Watch from midday until 4.00p.m., then

The crew's day generally started early and those not on watch would usually be woken by the boatswain's mates calling, 'All hands ahoy', at 7.00a.m. The men would dress quickly and lash up their hammocks in tight rolls, stored in the hammock nettings on the railings around the upper deck. Each man had two hammocks, one in use and the other clean. After a quick breakfast, those on watch returned to their duties and the others carried out their daily tasks.

Inevitably, mealtimes were significant highlights of the day. Breakfast usually consisted of a rough oatmeal gruel or porridge, known as 'burgoo', sweetened with molasses or possibly cocoa. For the purposes of preparing the midday meal the crew were divided into a number of individual messes, each consisting of around eight men. They would eat at

the First Dog Watch, the Last Dog Watch, First Watch, Middle Watch, Morning Watch and Forenoon Watch. Five of the watches were for four-hour periods and the two dog watches were just two hours each, which avoided one group getting the same watch every day.

narrow wooden tables, hooked up from the ceiling between the guns so they could swing with the pitch of the ship. The officers ate in their own wardroom at the back of the middle gun-deck, while the captain and admiral had their own quarters. Each mess appointed a mess cook for that week and he would collect the food and prepare it in time to be cooked in the galley. On *Victory* the galley is located on the middle gun-deck and is equipped with a cast-iron cooker known as a Brodie stove. This had two ovens that could bake up to 80lb (45kg) of bread, plus a pair of large copper kettles for making stews and a grill with an automatic spit for roasting meat. It is said that the only people allowed to whistle on board ship were the cooks, as a whistling cook couldn't spit in the food or eat it and whistle at the same time.

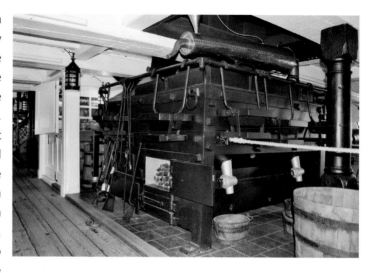

Dinner consisted of boiled meat or, if no fresh meat was available, there was 'junk', salted meat or fish stored in barrels which had to be soaked for several hours to remove the brine. The official ration was a pound or two of meat per week, but

▲ The cast-iron Brodie stove is situated beside the galley on the middle gun-deck.

pea flour and baked in the royal bakeries on shore. Much has been written about this notorious delicacy – mostly on the high protein content of weevils or maggots – probably best eaten in the poor light. Sometimes the evening meal would be supplemented with a 'duff', a mixture of flour and dried fruit.

By the general standards of the times, Nelson's sailors ate reasonably well. Just as well, as much of their work was very physical in nature. The Admiralty was also aware of the danger of scurvy caused by the lack of certain vitamins, especially vitamin C, and this was countered with fresh fruit when available and lime or lemon juice after that. Storing fresh water was also a problem, which is why the sailors had such a generous drink allowance, maybe four or even eight pints of beer

this was often more bone, fat and gristle than anything else. Fresh vegetables were only available while supplies lasted, and after that they might get dried peas, rice or oatmeal with the meat. One item that was supposed to keep well was the hard ship's biscuit, made of mixed wheat and

a day plus a pint of grog – water mixed with either rum or brandy to make it drinkable. When the beer ran out, they would be issued with wines or spirits. The men were accustomed to this amount of alcohol, although drunkenness on duty was a punishable offence. After their evening meal the Royal Marine drummer would 'beat to quarters', the signal for every man to attend his allocated position for battle. Then the hammocks would be brought down from the top deck.

Health was a constant concern aboard ship. If the diet did not kill you then there were a host of infectious diseases out there that could, especially when sailing in the tropics. On *Victory* the sick bay is located towards the front of the upper gun-deck, as far away from the main living areas as possible. During battle, the injured were

treated by the surgeon down in the orlop. As far as hygiene goes the men were required to keep themselves and their clothes clean as best they could. The ship's 'head', the toilets, were located in the bow just in front of the sick berth. These were

▲ *Simple wooden plates and bowls for the crew; their square plates gave rise to the expression 'a square meal'.*

➤ *Contemporary illustration of seamen eating at a table slung from the ceiling.*

simple seats with a hole open to the ocean. The captain and admiral both had their own private heads in their quarters, one on each side to be used according to the wind direction.

Another facet of living in such crowded and cramped conditions for months on

Berths for the sick, located at the forward end of the upper gun-deck to keep the risk of contagion away from the other men.

The ship's purser as depicted by Thomas Rowlandson, c.1800.

The head, in other words the toilet, located near the bows on the upper gun-deck.

end, if not years, is the delicate question of their sexual activities. The Victorians brushed over such nastiness, but Winston Churchill once famously summed up naval tradition as 'Nothing but rum, sodomy and the lash'. The validity of his claims are hard to substantiate. In the late eighteenth century homosexuality was considered to be a mortal sin, in theory punishable by death, and for the most part a ship's company reflected the mores of the day. As mentioned in 'Nelson's Navy', it is known that there were more women on board the ships than official records might suggest, and there is no shortage of accounts that prostitutes came aboard in droves whenever a ship put into port. Hence, we have the old adage that sailors have a 'wife' in every port.

◀ *The admiral's cot, a simple wooden box suspended from the ceiling.*

A SHIP OF WAR

Broadside to broadside our cannon balls did fly:
Like hailstones the small shot around our deck did lie.
Our masts and rigging was shot away,
Besides some thousands on that day
Were killed and wounded in the fray,
On board the man O'war.

Sea Shanty written after the Battle of Trafalgar

When she was new, HMS *Victory* carried thirty 42-pounder guns on the lower gun-deck, twenty-eight long 24-pounders on the middle deck, thirty 12-pounders on the upper gun-deck and twelve long 6-pounders on the top decks. This armament was constantly being changed, partly at the direction of her commanding officer and partly through the improvements made to naval gun power during her service life. At one time, after her great repair in 1803, she had thirty 32-pounders, twenty-eight 24-pounders, thirty 12-pounders, twelve 12-pounders and, on the forecastle, two 12-pounders and a pair of 32-pounder carronades. The latter had enormous smashing power but relatively short range, but that, at the end of the day, was what a ship-of-the-line was all about – smashing power!

In the age of sail a ship's guns were arranged on each side of the hull and could only fire in one direction. Firing all of the guns on one side was known as a 'broadside', and when they were fired on both sides it was a 'double broadside'. Because the guns were only accurate at short range, the opposing ships had to sail closer and closer to each other, vying for position and advantage, each hoping to fire the first broadside. The orthodox method of attack was 'in line ahead', with both fleets strung out in a line parallel to each other. Aiming a gun in the rolling sea generally depended on the attitude of the ship. The British preferred to take the 'weather gauge', with the wind blowing towards the enemy, causing their own ships to heel with the wind enabling them to fire on the down roll into the enemy hulls.

This tactic caused devastation among the enemy gun crews and explains why the French casualties at Trafalgar were many times higher than the British. The French, for their part, concentrated their fire on the

▲ *The upper gun-deck is equipped with thirty 12-pounders.*

masts and rigging, firing their guns on the upward roll to cripple the British ships.

Another less orthodox tactic, used to such devastating effect at Trafalgar, was to sail a column of ships directly into the enemy's line in order to fire upon their bow or stern, the weakest part of the ship. Known as raking fire, this was especially risky as it exposed the lead ship to the enemy's broadsides and was only undertaken by the boldest commanders.

The big 32-pounders seen on HMS *Victory* are long pattern Blomefield guns named after their designer, Sir Thomas Blomefield, who was Inspector General of Ordnance at the Woolwich Arsenal. Each one had a maximum range of 2,600yds (2,366m), but in practice they were fired at much closer range. The main part, consisting of the rounded breech and

INAL, DECK, OF TRAFALGAR, GUNS, H.M.S. VICTORY.

◀◀ Doing battle at sea usually involved bombarding the enemy at very close range.

◀ A postcard photograph from around 1905 showing four of the twelve Trafalgar guns which have been returned to Victory. *(CMcC)*

the cylindrical bore, was 124in (3.24m) long, including the cascable, the rounded iron ball at the rear end. On either side, extended arms called trunnions supported it on the wooden carriage.

Each gun was fitted with tackles, a system of ropes and pulleys, to move it out from its reload or recoil position or to train it left or right when aiming. Running back from the carriage is the train or preventer tackle, attached to a ring bolt in the deck

near the centreline of the ship. This stopped the gun from running out on its own due to the rolling of the ship, or it could be used to run the gun inwards if the slope of the ship was against it. As can be imagined, a gun of this size had an enormous recoil and a stout hemp rope, known as a breeching rope, was used to restrain it and hold it in position for reloading. This passed through the neck ring on the cascable, down through the strong iron ring bolts each side of the carriage, and secured to the ring bolts on either side of the gun port.

Loading and firing a gun followed a well-practised routine. The gun-powder cartridge, a cloth bag containing the gun-powder – a full charge usually containing 11lb (4.9kg) – was inserted by means of the wooden rammer. A wad of rope yarn was then rammed home, followed by the shot and then another wad to hold it in place while the gun was aimed. The gun captain would then take his priming-iron, an implement like a knitting needle, and thrust it down the touch-hole at the breech to clear the vent and cut into the cartridge.

◀ A display of the different types of shot, with a cut out of a loaded breech at the top.

The priming or quill tube, filled with a stiff paste made from finely milled powder and spirits of wine, was then inserted into the touch-hole and ignited with a slow-burning match. After 1780, the guns on some ships were fitted with a flint-lock mechanism operated by a lanyard. These were far safer and enabled the gunners to fire more rapidly.

The moment of firing was violent in every respect. The noise was deafening and the 2-ton gun would be thrown back to the limit of the breeching. It is said that a gun hot from repeated firing could leap from the deck after each shot, and many men were killed or wounded by the recoil of the guns. Before a fresh cartridge could be placed in the muzzle, a large corkscrew-shaped instrument called a worm was used to remove the fabric scraps from the

cartridge bases, usually after every fourth shot, to prevent the debris blocking up the vent hole. A sheepskin sponge was then thrust down to extinguish any smouldering embers, and the gun was ready for the next cartridge.

The shot fired by the guns and carronades took many forms, although the round-shot of cast iron was the most commonly used. The weight of the shot denoted the size of the gun, and a 32-pounder fired a 32lb (14.5kg) shot, for example. In close action, various other types of shot were used and some were designed specifically to cut the rigging and spars. Bar shot took the form of two balls joined by a solid iron bar, while chain shot comprised a pair of iron hemispheres linked by an iron chain. Both of these were capable of cutting a man's head clean off. Grape shot, used mainly against boats, consisted of a bundle of small iron balls, each weighing 2lb (0.91kg), enclosed within a canvas bag. A bag of sixteen balls was used with a 32-pounder. Upon firing, the bag would disintegrate and the balls would spread out with devastating effect.

H.M.S. Victory.   The way they worked the Guns at Trafalgar.

Rear-Admiral Nelson's Conflict with a Spanish Launch, July 3rd, 1797.

Another factor in Nelson's victory at Trafalgar was the greater speed with which the British guns were loaded and fired in comparison with the French and Spanish guns. During an engagement, the guns needed a constant supply of fresh cartridges which had to be brought up from the ship's magazine. It is often suggested that this task was carried out by the ship's boys who earned the title 'powder monkeys', but there were simply not enough boys on *Victory*. It was most probably done by teams of men, boys and women working in relay.

If you visit HMS *Victory* you will see twelve of the Napoleonic-era guns; nine 32-pounders on the lower gun-deck and three 24-pounders on the middle gun-deck.

May the great God, whom I worship, grant to my country and for the benefit of Europe in general, a great and glorious victory: and may no misconduct, in anyone, tarnish it: and may humanity after victory be the predominant feature in the British fleet.

Nelson's prayer before the Battle of Trafalgar

By 1801 both England and France were weary of war, and on 22 October their governments signed the Peace of Amiens treaty which brought a brief respite for both countries. Under Napoleon Bonaparte's direction the French used this suspension of hostilities to prepare for an invasion of England and orders were placed for the construction of a fleet of over 3,000 invasion craft. Meanwhile, the major refit of *Victory*, known as the 'great repair', continued. By the time of its completion in 1803 the storm clouds of war were gathering once more.

On 16 May 1803, *Victory* set sail from Portsmouth under a new captain, Thomas Hardy, and as Nelson's flagship. Their destination was Malta where Nelson would take command of a squadron of ships before joining the blockade of Toulon. In January the French fleet, under Admiral Pierre-Charles Villeneuve, had broken

Did you know?
If you are feeling groggy it is probably because you drank too much grog, a mixture of rum and water.

heading off for the West Indies. Nelson set off in hot pursuit, desperate for a good scrap. After arriving in the Caribbean he failed to locate the French ships, which were already making their way back to France. They were intercepted by a British fleet under Sir Robert Calder in the Battle of Cape Finisterre, but suffered only minor losses before making a bolt for the port of Cadiz. Dejected and frustrated by his fruitless chase back and forth across the Atlantic, Nelson return to London expecting a hostile reception. Instead, he was greeted as the man who saved the West Indies from a French invasion.

When news arrived that the French and Spanish fleets had combined forces and were at anchor in Cadiz, Nelson sailed from Portsmouth on 15 September 1805 to join the British fleet blockading the port, taking through the blockade but were forced back into Toulon by bad weather. Villeneuve tried again in April, and this time made it through the Straits of Gibraltar before

over from Rear-Admiral Collingwood. Now Admiral Villeneuve was in command of a sizeable combined Franco-Spanish fleet, but was reluctant to leave the port and engage the British who he regarded as a superior fighting force. Over the ensuing weeks a handful of British frigates kept an eye on the harbour while Nelson gathered his forces out of sight away from the shore. By the beginning of October the British ships desperately needed provisioning, and five ships-of-the-line were despatched to Gibraltar to collect supplies.

On 18 October 1805 Villeneuve's fleet was spotted by HMS *Euryalus*, making its way out of Cadiz. Nelson ordered his ships to turn towards the enemy, shadowing them at a distance to ensure he would engage them out at sea. At 4.00a.m. on the morning of 21 October, Nelson ordered

▲ *Victory's* starboard flank bristling with guns. The glazed windows are a later addition to keep rainwater out.

a change of direction towards the enemy. At 6.00a.m. he gave the order 'prepare for battle'.

Nelson's fleet comprised twenty-seven ships-of-the-line, including *Victory*, plus four second rates and twenty third rates, in addition to six smaller ships including

four frigates. They were outnumbered by Villeneuve who was fielding thirty-three ships-of-the-line plus a handful of frigates and brigs. The conventional tactic called for the opposing fleets to run parallel to each other with their ships in line ahead, and to blast each other with broadsides into submission. In practice, this method of attack was often indecisive and, anyway, Nelson had thrown the rule book out long before. He carefully planned an attack which would take the enemy by surprise and give the British the upper hand.

Two columns of British ships, one headed by Nelson in *Victory* and the other by Collingwood in the *Royal Sovereign*, would drive straight at the enemy's line cutting it in two. This would enable them to close with the Franco-Spanish ships as quickly as possible, bringing on a mêlée in which the faster gunnery skills of the British would come into their own, while isolating the van or front part of the enemy fleet from their flagship's signals. It was a high-risk strategy as it meant the leading British ships would be attacking head on, exposing their bows to raking fire from the French guns.

At 11.45a.m. Nelson instructed his signalman, Mr Pasco, to hoist the message, 'England confides that every man will do his duty'. Pasco replied that if he substituted 'expects' for 'confides' it would require fewer flags as confides had to be spelled out in full. 'That will do, Pasco, make it directly,' the Admiral replied.

With the Franco-Spanish fleet arranged in a ragged crescent running northwards, the two British columns approached from the west. As they closed, Nelson led his column into a feint towards the van of the enemy ships and then abruptly turned towards its centre. With light winds, the

British ships endured the raking fire for more than forty minutes. Collingwood, in the *Royal Sovereign*, broke into the line just astern of the Spanish flagship, *Santa Ana*, and suddenly the tables were turned with the British guns letting rip into her stern. This was the most vulnerable part of a ship

▲ *A 24-pounder in position on the middle gun-deck.*

and in the minute or so it took the *Royal Sovereign* to pass by her, guns loaded with double shots to maximise damage at close quarters ripped through the Spanish ship from end to end.

At 12.45p.m. *Victory* cut into the enemy line between Villeneuve's flagship, *Bucentaure*, and the *Redoubtable*, passing so close to the *Bucentaure* that they almost touched. *Victory's* guns burst into a thunderous roar, firing straight into the stern galleries of the French ship before coming alongside the *Redoubtable*. The two ships locked masts and the crew of the *Redoubtable* gathered ready to board the *Victory* while all about them the opposing ships engaged at close range. With the air thick with smoke and the smell of sulphur it was a truly hellish scene. Broadside after broadside sent a storm of shot ripping through the rigging or tearing through the oak hulls. It was a deliberate tactic of 'shock and awe', and the mathematics of slaughter favoured those who scored the most fatalities. And yet, at the heart of this killing zone, an extraordinary piece of theatre came into play with all the inevitability of a Greek tragedy. From the gunners and the powder monkeys all the way up to the commanding officer, each

man had his role to play. Each man did his duty. There was no place to hide, below decks or above, and the officers led by example: completely unprotected they paced the quarterdeck in full view of their men – and their enemy. This is why Nelson

Quarterdeck looking forward. A brass plaque marks the spot where Nelson is believed to have fallen.

Quarterdeck, looking forward.

wore his dress uniform that fateful day; not because he was vain but because he knew that his men had to see him facing the same dangers that they faced.

As every fibre of the ship shuddered with each broadside, the air thick with shot and wooden splinters as long as knives, a French soldier fired into the fog of battle.

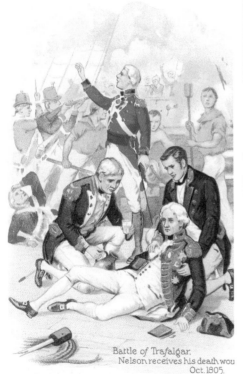

Battle of Trafalgar,
Nelson receives his death wou[nd]
Oct. 1805.

His musketball tore through the sails and rigging, passed through Nelson's shoulder sending a shock wave of destruction like a fist through his internal organs, smashing his spine and lodging in his lower back. Nelson spun to the side with the impact

▲ The spot where Nelson is thought to have died on the orlop deck.

◄ Hardy tends to Nelson on the quarterdeck moments after the fatal shot.

and then slumped to the floor. Captain Hardy, by his side, rushed to his aid. 'They have done for me Hardy,' he said. 'My backbone is shot though.' Having covered his medals with a cloth, lest the men should recognise who it was, they took him below decks down to the orlop where the surgeon treated the wounded. Fading in and out of consciousness for several hours, Nelson was desperate to learn of the battle's course. By 2.30p.m. Hardy came down to inform him that a number of enemy ships had surrendered. Nelson died at around 4.00p.m. Britain had proven its naval supremacy, but had lost its greatest hero.

We do not know whether we should mourn or rejoice. The country has gained the most splendid and decisive victory that has ever graced the naval annals of England; but it has been dearly purchased.

King George III on receiving the news of Nelson's death and the great victory at Trafalgar

On the evening of the battle, news of Nelson's death slowly filtered through the rest of the British fleet. Those that had not heard directly looked across to the *Victory* and saw that no commander-in-chief's night signal burned. The men mourned the loss of their beloved admiral bitterly, but their thoughts soon turned to more pressing needs. The ships of both sides had been left battered and bloodied by what had been the most brutal naval engagement in history. After the Battle of the Nile, the British officers had held a celebratory dinner, but there would be none of that after Trafalgar. The ships of the two navies were described as 'lying like logs on the water'. Several were completely dismasted, and all had been severely damaged. Thousands of men had died and countless others were being treated for their wounds. It is said that the decks were covered in splinters and debris, awash with blood and gore.

97

A tribute to Nelson at the old naval college at Greenwich.

With Nelson dead, Vice Admiral Collingwood assumed the command-in-chief and transferred his flag from the crippled *Royal Sovereign* to the frigate *Euryalis*. Before he died, Nelson had ordered that the fleet should anchor immediately after the battle to ride out the approaching storms, but fearful that the ships would be blown on to the treacherous Spanish coastline Collingwood prepared to take them to Gibraltar. All of them. Not only the British ships, but also the captured vessels of the enemy. These were regarded as legitimate prizes of war, and their considerable value would be shared out among the British officers and crews. A captain might receive enough to set him up for life. For a poorly paid seaman, their more modest share still represented a hefty chunk of money.

◄ *Drawing of the Battle of Trafalgar.*

⚓

Did you know?
There have been six ships called HMS *Victory*, but none have been given that name since 1765.

On board *Victory* the masts and rigging were in a horrific mess, and Captain Hardy ordered his men to make emergency repairs as best they could. He had Nelson's body placed in brandy, in order to preserve it, inside a large water butt known as a 'leaguer' which was lashed to the mainmast in the middle gun-deck and placed under the protection of an armed marine. There was no way *Victory* could get to Gibraltar

under her own sail, so she was taken under tow behind the *Neptune*. For over a week, Collingwood's rag-bag fleet of around fifty vessels fought against the worsening storms. Towing the hulks was especially difficult as they rolled heavily in the swell, like 'hogs' it was said, without the steadying effect of the masts. Collisions with the towing vessels were frequent. On *Victory*, the huge main yard was torn away in the gales, dragging rigging and sail with it. In many cases tow ropes gave way and the ships were set adrift to be driven against the rocks. Eventually Collingwood had to cut his losses and ordered the captured vessels to be set lose to save their own ships. For the British sailors it was not just the loss of their prizes that concerned them, for there was a genuine sense of humanity for the French and Spanish prisoners who were on board these ships, and there are many accounts of their heroic actions in saving as many lives as they could.

Victory arrived at Gibraltar on 28 October 1805, and the remaining dead sailors were buried in the Trafalgar Cemetery on the island. After further makeshift repairs *Victory* set sail for England, arriving at Portsmouth on 5 December. Nelson's body was transferred to the yacht *Chatham* and taken to Greenwich, where it lay in state for three days. Meanwhile, Captain Hardy took *Victory* up the River Medway and to her birthplace at Chatham Dockyard.

In London, plans were made for a state funeral for Nelson, a rare concession for a non-royal and one regarded by some critics as a cynical ploy to bolster the King's flagging popularity. Not content with one coffin, Nelson was treated to four. The

◀ *Thomas Hardy, Nelson's close friend and captain of the* Victory *at Trafalgar. (CMcC)*

inner one was of wood, from the mainmast of *L'Orient*, a French ship destroyed in the Battle of the Nile, then cased in lead with an outer wooden coffin, and the whole lot

➤ More men died in the storms afterwards than died in the battle itself.

Did you know?

It cost £63,176 to build *Victory* in 1765. That is roughly the equivalent of building an aircraft carrier nowadays.

The King's cutter, used to carry Nelson's coffin along the Thames, is now displayed at the National Museum of the Royal Navy in Portsmouth.

placed inside a gilt casket. It is said that they had to break his bones to get his body out of the leaguer and into the coffin.

With sixty boats in attendance, his body was carried by barge from Greenwich to Whitehall on what was described as the

Did you know?

Victory could hoist a maximum of thirty-seven sails.

in St Paul's Cathedral on the following morning, 9 January 1806. Thousands lined the streets to see the procession which was so long that its head had reached St Paul's before the rear end had got going, and great respect was paid to the sailors from Trafalgar. Some of the seamen carried the damaged flag flown on *Victory*, and several were seen to tear small strips off to place in pockets near their hearts. Nelson was finally laid to rest in the crypt of St Paul's in a large black sarcophagus, originally made for Cardinal Wolsey three centuries earlier. He lies immediately under the centre of the great dome.

In April 1806 most of the crew of HMS *Victory*, by then undergoing a refit, were transferred to HMS *Ocean*, a new ship which became Collingwood's flagship. An established ship's company, especially

'greatest aquatic procession ever beheld on the River Thames'. It was then placed in the Admiralty overnight before the funeral

◄ 'A return from an invasion, or Napoleon at a nonplus.' This caricatures shows a dejected Napoleon coming ashore at Calais to face the ridicule of fisherwomen, a French soldier and an English sailor, with the British fleet in the background.

the *Victory*'s, was highly prized, and the process known as 'turning over' was not uncommon. *Victory* was put in ordinary, in reserve, and refitted as a second-rate ship. That same year, Napoleon signed a treaty with the Russian Tsar Alexander which

Here Nelson Fell.

➤ *King Edward VII visits* Victory *and examines the spot where Nelson fell, probably in 1905 to mark the centenary of Trafalgar.*

aimed to exclude British merchantmen and warships from the Baltic. *Victory* became the flagship of Admiral Sir James Saumarez, who was tasked with keeping the Baltic open to British shipping as well as supporting Sweden, which was one of the few states in the region to oppose the French and Russians. The Baltic remained important to the Admiralty for many reasons – it was a vital source of hemp, flax and pine. By 1812 the situation in the Baltic had eased and *Victory*'s active career ended on 7 November 1812 when she was moored in Portsmouth Harbour.

The absurdity of the present rig and appearance of the *Victory* is little short of an insult... It is incredible that she should so long have been allowed to remain in this plight, for to preserve a vessel on purely sentimental grounds and then mutilate and disguise her in every possible way is imbecile.

Frank H. Mason – *The Book of British Ships*, **1911**

Nelson's victory and dramatic death at Trafalgar was marked by a veritable forest of towers and columns, plus countless monuments, springing up in public squares and parks the length and breadth of the nation. But what of his equally famous flagship?

HMS *Victory*'s active career officially ended on 7 November 1812, after which she was moored in Portsmouth Harbour to be used as a depot ship and to await her fate. The story goes that she only escaped scrapping through the intervention of Thomas Hardy, First Sea Lord and former captain of *Victory* at the time of Trafalgar, who tore up the order. Forgotten and forlorn, few people took the trouble to visit the ship. Interest was revived in 1833 when the young Princess Victoria, then aged fourteen, persuaded her mother, the Duchess of Kent, into taking her to see *Victory*. Apparently she had been

➤ *In 1889 Victory was fitted as a Naval School of Telegraphy and continued in this role until 1904.*

➤➤ *1905 was the centenary of the Battle of Trafalgar, not just Nelson's death, but perhaps celebrating the battle was not appropriate in restoring the entente cordiale between former enemies. (CMcC)*

reading about the life of Nelson and was much amused by her visit to Portsmouth. This started something of a fashion, but nothing on an organised level, and visitors were rowed out to the ship in small groups.

▶ Victory *in the No.2 Dry Dock at Portsmouth, resting on a cradle of steel.*

Did you know?

Victory was built with the wood from around 6,000 trees, but the ship you see on display at Portsmouth is only about 30 per cent original.

In 1889, *Victory* was fitted as a Naval School of Telegraphy. Instruments were set up in the admiral's after cabin, and over a two-month course the signal ratings came to grips with semaphore signalling, both mechanical and manual, flag hoisting –

using the fore and mizzenmast to represent different ships – plus flashing signals and, later on, telegraphy. In 1904, the school was transferred temporarily to HMS *Hercules*, and two years later it was moved to a shore-based establishment. *Victory* was abandoned to grow old disgracefully and it was only when, in 1903, she was accidentally rammed by HMS *Neptune*, which was being towed on its way to the scrappers yard, that everyone suddenly realised how fragile *Victory* had become. Emergency repairs were made to the severely damaged hull, and it was only the personal intervention of King Edward VII that saved her once again.

By the time of the first centenary of the Battle of Trafalgar, in 1905, there was a renewed groundswell of interest in the ship as an emblem of Britain's

◄ Victory *at Portsmouth in the 1930s, from* Shipping Wonders of the World. *(CMcC)*

role as a great sea power. During the centenary celebrations a submarine was moored alongside to supply electricity

111

➤ *Crowds of visitors descending on the dockyard at Portsmouth to see* Victory *and a group of British submarines, c.1930.*

for lights strung throughout her rigging. Then, in 1910, the Society for Nautical Research was formed with the intention of preserving *Victory*. Unfortunately, the Admiralty showed little interest as it was more concerned with developing new ships in an escalating arms race leading up to the First World War.

Neglected and no longer serving a useful purpose, the once-proud flagship faced a bleak future of steady deterioration. If nothing was done to halt the decay she would be good for nothing but firewood. Thankfully, there were some people who cared passionately for the old ship and her plight did not go entirely unnoticed. The much-respected writer Frank H. Mason, who also happened to be an accomplished maritime and railway artist and poster designer, let rip with a tirade on the state

⚓

Did you know?
If laid in one long line, the cordage used to rig *Victory* would stretch to approximately 26 miles (41.9km).

of *Victory* in *The Book of British Ships*, published in 1911. One particular gripe was the state of her masts and rigging, which had been drastically cut down to reduce strain on the hull. Instead of being 'extremely interesting and beautiful', he

real possibility that she might sink at her moorings. A public campaign was started to 'Save the *Victory*'. With support from the wealthy shipowner, Sir James Caird, the Society for Nautical Research raised £50,000 which spurred the Government and the Admiralty into action. On 12 January 1922, *Victory* was moved into her new permanent home in No.2 Dry Dock at Portsmouth Dockyard, supported by iron cradles standing on a concrete plinth, where work began on restoring the ship to her Trafalgar appearance. In 1928, King George V unveiled a tablet to celebrate its completion.

described it as 'little short of an insult' to the traditions of the old order of the sea.

Fortunately, his was not a lone voice. The First World War may have put such concerns on hold, but by 1921 she was in such a poor condition that there was a

In 1941 a Luftwaffe bomb fell between the dock and the foremast and exploded underneath, blowing out one of the steel cradles and damaging the foremast. German propaganda was swift to claim that this symbol of British pride had been destroyed, and in response the Admiralty assured the public that this was definitely not the case.

After the Second World War, the second phase of *Victory*'s restoration focussed on replacing much of the timber which had become rotten. Remedial measures included putting an end to the navy's practice of washing down the decks with water. A number of holes were cut at the bottom of the hull and several bulkheads were removed to improve ventilation. In addition to the threat from fungal rot, the timber was also under attack from the dreaded deathwatch beetle, and since the 1950s the ship has been regularly fumigated. Since the 1960s, major restoration work has seen much of the original oak replaced with Burmese teak and African iroko. Both

of these hard woods are oily and, therefore, more resistant to pests and rot than the oak.

HMS *Victory* is the oldest naval ship in commission and serves as the flagship of the Second Sea Lord in his role as Commander-in-Chief of the Royal Navy's

▲ *In 2010 a major restoration project was started at Portsmouth to replace some of the hull's timber.*

Launched in 1860, the iron-clad HMS Warrior *is the only surviving member of Queen Victoria's Black Battle Fleet.*

Did you know?

Apart from Nelson, *Victory* had been the flagship for fifteen different admirals before the Battle of Trafalgar, and over seventy more since then.

Home Command. *Victory* is listed as part of the National Historic Fleet, Core Collection, and as a museum ship she attracts around 350,000 visitors per year. Work on maintaining her in best condition is ongoing. In 2010, some sections of the ship were shrouded in scaffolding once more as part of a major restoration programme to make significant repairs to the hull planking. Great emphasis is being placed not just on preserving the fabric of the ship, but also on the contextualisation of life on board. Today's visitors will see replicas of many items on display such as the bowls, beakers and tankards used by the men, and the drapes made by the officers' wives to make their wooden sleeping cots seem less like coffins, artefacts which help to bring this great ship, and its remarkable story, alive.

POSTSCRIPT – FROM OAK TO IRON

In 1803 Admiral Horatio Nelson toured the Royal Forest of Dean, Gloucestershire, to investigate for himself the availability of timber for the Admiralty. He was appalled to discover that huge swathes of the forest had been plundered for the charcoal-burning industry and he immediately recommended to parliament that there should be a replanting programme to meet their future needs. In 1808, three years after Nelson's death, the replanting began. But what Nelson failed to realise, what he could not possibly have guessed, is that a revolution in shipbuilding would see oak replaced by iron before the saplings had grown into usable trees.

Ironically, the seeds of this revolution from oak to iron can be found in Portsmouth, a few streets away from the entrance to the Historic Dockyard. The engineer Isambard Kingdom Brunel was born at No.1 Britain Street on 9 April 1806, less than six months after the Battle of Trafalgar. It was Brunel who built the world's first large-scale iron ship, the SS *Great Britain*, which was launched in Bristol in 1843. And if you travel to Portsmouth to see HMS *Victory*, the first thing you will notice is another iron ship, HMS *Warrior*, moored near the dockyard entrance.

Launched in 1860, *Warrior* is magnificent, the sole surviving iron-clad ship from Queen Victoria's Black Battle Fleet. Amazingly, apart from the obvious difference in materials, Nelson's men would have had little difficulty in finding their way about this ship. So much would have been familiar to them from the rigging, as sail was used to supplement the steam engines, to the gun-decks with their rows of guns running down her flanks. *Warrior* is twice as big as *Victory*, but had a crew of around 750. Even so, she was many times more powerful, such was the rate of change.

As a footnote to this story, the saplings planted after Parliament's ruling in 1808 are now great oaks, and when timber was needed for the restoration of *Victory* in time for the bicentenary of Trafalgar in 2005, it was taken from the Forest of Dean, which has the largest area of mature oak trees anywhere in Britain. The locals have a name for them; they call them the 'Nelson Oaks'.

1744	The previous first-rate *Victory* is lost with all hands.
1758	14 July: A new ship-of-the-line ordered by the Admiralty.
	29 September: Horatio Nelson born at Burnham, Norfolk.
	December: Commissioner of Chatham Dockyard instructed to prepare a dry dock for the construction of a new first-rate ship.
1759	23 July: Keel laid in the Old Single Dock (renamed No.2 Dock).
1760	The name, HMS *Victory*, is chosen. She will be the sixth ship to bear the name.
1764	An eight-gun schooner, also named HMS *Victory*, launched. She serves in Canada before being burned in 1768.
1765	7 May: *Victory* is launched at Chatham Dockyard.
1778	May: *Victory* is commissioned.
	23 July: First Battle of Ushant.
1780	*Victory*'s hull sheathed below the waterline with copper to protect it from shipworm.
1781	12 December: Second Battle of Ushant.
1797	14 February: Battle of St Vincent.
1798	*Victory* was ordered to be converted into a hospital ship to hold wounded prisoners of war.
1799	Following the loss of HMS *Impregnable*, the Admiralty decides to recondition *Victory*.

1800	Work starts on reconditioning *Victory* as a first-rate ship. Extra gun ports added and paint scheme changed to the black and yellow pattern known as 'Nelson chequer'.
1803	11 April: Work on *Victory* completed, and on 14 May she leaves for Portsmouth. 16 May: Vice-Admiral Nelson hoists his flag on *Victory*.
1805	21 October: Battle of Trafalgar and death of Nelson.
1806	6 January: Nelson buried in St Paul's Cathedral.
1808	*Victory* helps to evacuate British troops from Corunna.
1812	7 November: End of *Victory*'s active career, moored in Portsmouth.
1825	HMS *Victory* becomes flagship of the Port Admiral.
1889	The ship is fitted up as a Naval School of Telegraphy, and remains a Signal School until 1904.
1903	The ship is accidentally rammed by HMS *Neptune*.
1905	100th Anniversary of Battle of Trafalgar.
1921	Save the *Victory* Fund launched.
1922	12 January: *Victory* put into dry dock in Portsmouth.
1928	Ship opened to the public.
1941	*Victory*'s hull is damaged by a Luftwaffe bomb.
2005	200th Anniversary of the Battle of Trafalgar.

APPENDIX 2 – PLACES TO VISIT

A selection of places to visit with either Nelson or HMS *Victory* connections:

HMS *Victory* and the Portsmouth Historic Dockyard

Hampshire: The obvious place to start. See not only HMS *Victory* and the Trafalgar sail, but also the National Museum of the Royal Navy, the *Mary Rose* Museum, plus the 1860 HMS *Warrior*, the world's first iron-hulled battleship.
www.historicdockyard.co.uk

The Historic Dockyard Chatham

Kent: An 80-acre museum site, this is where *Victory* was built in the Old Single Dock, renamed as No.2 Dock and since 2005 as the *Victory* Dock. It is currently occupied by HMS *Cavalier*, one of three warships on display.
www.thedockyard.co.uk

The National Maritime Museum

Greenwich, London: The world's largest maritime museum has Nelson relics, paintings, a large Nelson library and archive.
www.nmm.ac.uk

Nelson Museum, Monmouth

Monmouth, Wales: Nelson collection donated by the mother of Charles Rolls.

www.nelson-museum.wales.info

Nelson Museum

South Quay, Great Yarmouth, Norfolk: Displays on all aspects of Nelson's life.

www.nelson-museum.co.uk

St Paul's Cathedral

London: Wren's magnificent cathedral houses Nelson's tomb and the Flaxman memorial statue.

www.stpauls.co.uk

MEMORIALS AND MONUMENTS

Nelson's Column

Trafalgar Square, London: The big one, a 151ft (46m) column topped with a 18ft (5.5m) statue of Nelson looking towards the Admiralty. Designed by the architect William Railton and erected between 1840 and 1843. Landseer's famous lions were added at its base in 1867.

The 18ft (5.5m) statue of Nelson atop the column in Trafalgar Square was created by sculptor E.H. Bailey. Nelson is said to look in the direction of the Admiralty and Portsmouth.

Nelson's Monument

Bullring, Birmingham: A fine bronze statue by Richard Westmacott, 1809.

Nelson's Monument

Carlton Hill, Edinburgh, Scotland: Tower built between 1807 and 1815.

Nelson's Tower

Cluny Hill, Forres, Moray, Scotland: Octagonal tower erected in 1806.

Nelson's Pillar

O'Connell Street, Dublin, Ireland: Stone column destroyed by the IRA in 1966, the site is now occupied by the 394ft (120m) Spire of Dublin.

Nelson's Column

Place Jacques-Cartier, Montreal, Canada: Stone column with statue, erected in 1809.

Nelson's Statue

Heroes Square (formerly Trafalgar Square), Barbados: 1805 statue.

▲ Supposedly modelled on a telescope stood on its end, the Nelson Monument in Edinburgh. (StarsBlazkova)

Able seaman	An experienced sailor, usually at least twenty years old with five years' experience at sea.
Binnacle	Positioned just in front of the ship's wheel, it contains the compass.
Blocks	Wooden pulley blocks – *Victory* had 768 on the running rigging.
Boatswain	Warranted officer responsible for sails, rigging and rope.
Bowsprit	Mast extending from the bow.
Capstans	Rotating mechanism used to raise the anchors
Carronade	A short-range powerful gun which could be mounted on the top deck.
Chain shot	Projectile consisting of iron balls linked by a chain, used to cut rigging.
Commander	Captain of a sixth-rate ship.
Commodore	A probationary rank between captain and admiral.
Coxswain	Rating in charge of the captain's boat and crew.
Cutter	18ft ship-to-shore work boat.
First-rate ship	Naval ships were rated by the number of guns they carried. To be a first-rate ship *Victory* had to carry at least 100 guns.
Flag officer	Collective term for admirals and commodores.
Flotilla	A small fleet of ships.

◄ Edward Landseer's famous lions in Trafalgar Square were added in 1867. (US Library of Congress)

◄◄ Richard Westmacott's statue of Nelson was erected in Birmingham city centre in 1809.

Figurehead from HMS Nelson at Portsmouth Historic Dockyard.

Foc'sle	It was from the forecastle, near the bow, that the main and fore masts were controlled.
Foremast	First of the main set of three masts.
Frigate	A light fast vessel used mainly for intelligence gathering and scouting.
Grand magazine	Main store for gunpowder.
Grape shot	A cluster of smaller iron balls.
Guns	The ship's cannon.
Hold	Large storage area located in the lowest level of the hull.
Main mast	The middle of the three vertical masts.
Mess deck	Living area for the majority of the crew.
Midshipman	Young gentleman aspiring to become an officer.
Mizzenmast	The third vertical mast, at the stern of the ship.
Ordinary seaman	A useful seaman, but not an expert sailor.
Orlop	Storage and living area below the waterline.
Poop deck	Raised deck at the stern, used mainly as a viewing or signalling platform.
Powder monkeys	Sometimes young boys, they ensured a continuous supply of powder cartridges and shot.
Quarterdeck	The nerve centre of the ship, just in front of the poop deck.

◁ In dry dock alongside Victory is Monitor 33, the only British warship to survive from the First World War, resplendent in its black and white 'dazzle' camouflage.

Rammer	A flat-headed wooden stave used to push the cartridges and shot to the bottom of the guns.
Round shot	Cast-iron 'cannonballs'.
Running rigging	Used to hoist and adjust the sails.
Side tackle	Ropes and blocks rigged to return a gun from its recoil position, or to train it left or right.
Squadron	A group of ships under the command of a flag officer.
Standing rigging	Rigging that supports the masts and bowsprit.
Tar	Jack Tar was the colloquial term for a seaman.
Tops	A platform where the top mast is attached to the lower mast.
Wardroom	Living quarters for the officers at the stern of the ship.
Warrant officer	Officers appointed by warrant to carry out specialist duties, such as surgeon, carpenter etc.
Yards	Cross pieces from which the square sails are hung.

THE BLUE FUNNEL LINE

A Portrait in Photographs and Old Picture Postcards

by

Terry Moore

Foreword by Hugh M. Wylie

S. B. Publications
1989

This book is dedicated to my wife, Valda.

First published in 1989 by S. B. Publications

5, Queen Margaret's Road, Loggerheads, Nr. Market Drayton, Shropshire, TF9 4EP.

British Library Cataloguing in Publication Data
Moore, Terry
    The Blue Funnel Line: a portrait in photographs and old picture postcards.
    1.   (Metropolitan County) Merseyside. Liverpool. Shipping Services. Blue Funnel Line.
    1865–1970.
    I. Title
    387.5'065'42753

ISBN 1 870708 12 1

Distributed in the U.K. by Macmillan Distribution Ltd.,
Houndmills Estate, Basingstoke, Hampshire RG21 2XS

Typeset and printed by Geo. R. Reeve Ltd., Wymondham, Norfolk NR18 0BD.

CONTENTS

CONTENTS CONTINUED

CONTENTS CONTINUED

CONTENTS CONTINUED

FOREWORD

by
Hugh M. Wylie
Late Publicity and Advertising Manager Ocean Steam Ship Co. Ltd., Liverpool

The Blue Funnel Line, like many of the other large passenger carrying shipping companies, used coloured postcards as an ideal method of advertising their passenger ships. No expense seemed to have been spared in the early days and many famous artists were used, including Kenneth Shoesmith R.A. and Maurice Randall, whose two famous paintings of "Aeneas steaming ahead" and "Sailor heaving the lead" were both portrait designs and very unusual for postcards.

Most of the early postcards were based on the paintings of Norman Wilkinson, James Mann and Sam Brown, the latter becoming famous in 1927 when he introduced a publicity design for a postcard which combined a young lady in a bathing costume sitting on the beach, looking out to sea at a Blue Funnel ship. However, he did not take the full credit for all of the picture, as a well-known figure artist added the lady. (See page 51).

In the early 1930s, Walter Thomas R.C.A. was responsible for most of the company's art work. He started his career as a naval architect and worked also for the Liverpool Printing and Stationery Co. Ltd., who subsequently printed all his works. However, after World War 2, he became a freelance artist and was one of the few marine artists whose designs never had to be altered. He was also one of the most successful artists to paint Blue Funnel ships.

Early in 1960, Peter Hatch, a London consultant and designer, was responsible for the company's art work. In 1963, Laurence Dunn did most of the work used in connection with publicising the "Centaur" which was specially designed for the Singapore – West Australia service carrying 200 first-class passengers as well as 2,000 sheep or 700 cattle.

Finally, the "Priam" class was painted by John Stobart and Laurence Dunn and a few postcards were also published using coloured negatives.

I wish this book every success and may it bring back many memories of a great shipping company which served its country and Liverpool so well.

INTRODUCTION

This small book cannot claim to be, nor was it written as, a definitive work on the history of the Blue Funnel Line and its classic fleet of ships. It is but a collection of photographs and postcards, mostly from my own collection, representing some of the different classes, or series of ships, which were regarded with admiration and affection wherever they went about their business.

Postcards were used by shipping companies as publicity for their passenger routes. Leading marine artists of the day were engaged to paint the various ships and these paintings were then published as postcards. This selection of official Blue Funnel cards, postcards from other sources and photographs have been arranged chronologically to cover a period of 100 years, beginning with the first ship, Agamemnon of 1866. Not every class is represented for pictures of ships from the last century are difficult to trace and a few from the early part of the present century are absent. However, those that are here portray the evolution of the cargo-liner from simple beginnings to a work of superb marine architecture. Comparing the early ships with the purpose-built ships of today, it is difficult to realise that they, too, were built for commerce and to earn profitable returns for their owners. The selection of illustrations end with the Priam Class of 1966, for with their sale from the fleet some twelve years later, the Blue Funnel Line was little more than a shadow of its former self, its identity eventually becoming lost in a group with very diversified interests.

Postcards are now very collectable, with some commanding extremely high prices. The collector is faced with the problem of not really knowing how many any particular company produced. This adds to the enjoyment of the search, only surpassed by the pleasure of finding cards that were not known to exist.

Photographs, too, are becoming much sort after by researchers and collectors who place a continuous demand for them on marine photographers.

As for the ships themselves, they are but a memory the like of which will never be seen again, so I would regard this book as one of nostalgia. A look back to not so long ago when they were admired for their graceful lines, dignity and bearing — worthy ambassadors of Britain's Merchant Navy that are no more.

Terry Moore,
King's Lynn,
March 1989.

A SHORT HISTORY OF THE BLUE FUNNEL LINE
(Alfred Holt & Co.)

Ocean Steam Ship Co. Ltd., China Mutual Steam Navigation Co. Ltd.
Nederlandsche Stoomvaart Maatschappij, Ocean N.V.
('N.S.M.O.')

The middle of the last century was a time of great industrial change and upheaval, caused by the use of steam-power to fuel the industrial revolution that was gathering apace in Britain. It was at this time, in 1865, that two Liverpool brothers, Alfred and Philip Holt, launched a shipping company that would become one of the most respected and individual of the world's merchant fleets.

Alfred Holt trained as an engineer, being apprenticed to the Liverpool and Manchester Railway Company. Shortly after serving his time, he entered the shipping business, though first becoming a consulting engineer in 1852, engaged in marine as well as locomotive engineering. He was joined by his younger brother, Philip, whose business sense matched the engineering skills of Alfred. Eventually, they accrued enough capital to register the Ocean Steam Ship Company on 11th January 1865.

The general opinion of the day was that the steamship could not compete with the fast sailing ships over the world's long-distance routes, owing to the large amounts of coal that would have to be carried, thus cutting down cargo space. Alfred was convinced that a propeller-driven iron ship with compound-engine was feasible and would pay her way. He had formulated ideas regarding compounding during the time spent working on locomotives, which he had experimented with and developed during his engineering consultancy.

The brothers had three ships built for their new company by Scotts of Greenock: 'Agamemnon', 'Ajax' and 'Achilles', all three sailing for China during 1866. The ships proved Alfred's theories to be well founded, but to be successful they had to find employment in the China trade in order to make a profit. This was achieved by conducting their business, in conjunction with the merchants of the day, in an orderly, honourable and reliable manner, together with the continuous improvement in ship design and efficiency. But an event occurred in 1869 outside their control which, whilst enabling them to gain the final advantage over the clipper ship, also brought greatly increased competition from rival shipping companies with less efficient vessels; this was the opening of the Suez Canal.

From these beginnings the fleet grew, with improvements to successive ships as they were built, based upon experience gained on the trade routes and the cargoes handled. Competition was fierce, so much so that Holts did not have everything their own way, as other shipping lines entered the lucrative Far East trade. In particular, the Dutch had formed a company of their own, operating a feeder-service throughout the East Indies bringing goods to Java for trans-shipment direct to European ports in their own ocean-going ships, without the need to use Holt's feeder-service via Singapore. Consequently, a new company was formed in Amsterdam in 1891 to counter the threat from the Dutch and to strengthen their own position, called Nederlandsche Stoomvaart Maatschappij Ocean N.V., a number of the older Blue Funnel ships were transferred to it under the Dutch flag, enabling them to compete on equal terms between Java and the European ports of Amsterdam, London and Liverpool. The 'N.S.M.O.' became a well-respected and important part of Blue Funnel with frequent transfers of ships between it and the parent company whenever the need arose.

As early as 1875, the Company had taken the decision to carry the whole of the risk on its ships, thereby aleviating the need to pay premiums for external insurance.

A service was introduced between Singapore and Java to Fremantle, Western Australia in 1890, followed in 1901 by new direct monthly sailings from Glasgow to Australia, via the Cape of Good Hope outwards and homewards by the Suez Canal, and further strengthened in 1910 by the addition of three passenger ships, 'Aeneas'. 'Ascanius' and 'Anchises' servicing the same route. The well-known 'Nestor' and 'Ulysses' followed in 1913.

During 1902 Holts were in a position to purchase a controlling interest in the China Mutual Steam Navigation Company, which had long been a serious competitor, with sailings from Liverpool to China. Not only was this competition eliminated, but Holts expanded their interests by adding new routes from Japan across the Pacific to the west coast of Canada and America, together with new cargoes for Europe, China, Japan and the Philippines. The purchase also brought thirteen ships, all under ten years old, into the fleet.

At the outbreak of World War I, Holts were a formidable enterprise with a fleet of 83 ships and possessed a very strong position in the Far Eastern, Pacific and Australian trades. However, Blue Funnel ships were well suited for the transportation of war material, being equipped with numerous derricks, they were self-reliant for the discharging of cargo. They were capable of above average speed, with large holds and exceptionally well-built, even to above the standards of Lloyds or any other classification society. Consequently, the requirements to which they were built became known as 'Holt's Standard' for nothing but the best was acceptable. Seventy-nine of the ships were requisitioned for war service and losses were heavy, 16 vessels being sunk and a further 29 badly damaged. In order to replace some of these ships, two small companies were purchased as the possibility of obtaining new tonnage was practically non-existent. The Indra Line, with seven refrigerated ships and routes from New York to China, was purchased in 1915, and the Knight Line with four ships in 1917. With the Panama Canal opening in 1915, Blue Funnel's routes would henceforth encircle the globe.

The years following the war were ones of reconstruction and modernisation of the fleet. Several Blue Funnel ships were provided with temporary passenger accommodation, at the request of the Government, to help satisfy the serious shortage that existed to the Far East. This proved to be so successful that four 11,300-ton steamers, known as the 'Sarpedon Class', were built with accommodation for approximately 140 passengers. Turbine-driven ships with both single and double-reduction gearing were added to the fleet also. Several ships were fitted with superheating and the Company's first motor-ship was built in 1923, followed in 1924 by the 'Dolius', with a Scott-Still engine. The modernisation and expansion of the fleet continued throughout the inter-war years, the solidarity of the Company enabling it to ride the Depression of the 1930s in relative safety.

In 1935 Blue Funnel purchased the Glen Line, a company that had always been a great rival — even in the days of the China tea trade. Holts immediately put in hand a building programme for their new acquisition consisting of eight twin-screw motor-ships of 9,000 G.R.T., known as the 'Glenearn Class', which would be the finest cargo-liners afloat. However, only three ships were delivered before the beginning of World War 2. The devastating toll during the war was 41 ships lost from the fleet of 88. In order to maintain its prominent position on the world's trade routes, the Company replaced its losses with six 'Victory' ships, eight 'Liberties' and three British-built standard ships.

In January 1947, the first of the 'A' Class was delivered; the class providing 27 ships built between 1947 and 1958. Also included in the reconstruction of the fleet, were eight fast 10,000 G.R.T. steam-ships with accommodation for 29 passengers; four 'H' Class ships for the Australian route, also provided with refrigerated space; and four 'P' Class ships for the Far East. Passenger ships to Australia had always called at Capetown but this service was discontinued in 1956, routed thereafter by way of the Suez Canal and Aden.

Blue Funnel Ships had long been associated with the carriage of pilgrims to and from the Far East and Jeddah, each ship capable of taking 1,000 or more pilgrims who were accommodated in the 'tweendecks' and also as deck passengers. In November 1958, Blue Funnel bought an ex-German liner, built in 1936, that had been taken over as a prize in 1945 and subsequently used as a troopship under the names of 'Empire Doon' and 'Empire Orwell'. Renamed 'Gunung Djati', she was converted to a pilgrim-carrier with accommodation for 106 first-class and 2,000 unberthed pilgrims.

The six ships of the 'M' Class eventually superseded the 'A' Class, being faster and larger vessels with their bridge superstructure and funnel placed slightly aft. Further changes were made by the Company when, in 1964, the 'Centaur' was introduced for the Singapore – West Australia service; the first Blue Funnel ship to be built with her superstructure aft and a newly-designed tapered funnel.

In 1966 saw the introduction of the 'Priam' Class: eight ships built for the express cargo-liner service to the Far East; four serving Blue Funnel and four serving the Glen Line. These ships possessed a completely new design with cranes replacing the traditional derricks, a Stulcken heavy-lift derrick added amidships, and the superstructure topped by a tapered funnel situated aft. However, after only twelve years of service, the rise of the container ship made them outdated and they were gradually sold out of the fleet.

Changes were also taking place in the structure of the Company itself. In 1965, Ocean Steam Ship Co. became a major shareholder in Overseas Containers Ltd. and then merged with the Liner Holdings Group which consisted of the Elder Dempster Line, the Henderson Line and the Guinea Gulf Line. This new grouping was named Ocean Fleets Ltd. The integration saw Blue Funnel ships sailing to West Africa and Elder Dempster vessels to the Far East. The name Alfred Holt disappeared the same year.

In 1969, Ocean Fleets joined with P&O to form Pan Ocean Shipping and Trading Company, and later in 1972, the Cory company joined the Group. These changes led to further restructuring and the Group changed its name to Ocean Transport and Trading — the Blue Funnel Line existing in name only. The operating divisions included: Ocean Titan (tankers and bulk carriers); Ocean Fleets (Blue Funnel, Glen Line and Elder Dempster); and Ocean Cory (towing, storage and distribution). During the 1970s, many changes were taking place to traditional trading routes, and in 1978, 'Protesilaus' made the last conventional sailing from Liverpool to the Far East. The last Glen Line vessel was sold the same year, and in 1980, the last Elder Dempster ship sailed for West Africa.

More and more ships were sold off during the late 1970s and early 1980s, with the remaining ships employed on the West African trade routes and round-the-world cargo services. The smaller fleet reflecting the demise of the British shipping industry.

In 1986, the Group finished its involvement with Overseas Containers Ltd., when the latter became a wholly-owned subsidiary of P & O. Finally, in December 1988, the Ocean Group sold off its last two ships and withdrew from deep-sea shipping operations completely.

The only maritime interests which remain for the Ocean Group are its involvement with Cory Towing and Ocean Inchcape (part of the offshore oil industry).

AGAMEMNON, 1865.

1865 by Scott and Co., Greenock for Ocean Steam Ship Co. 2,280 G.R.T. 310′ x 39′. Compound steam-engines, 945 I.H.P. Single-screw, 10 knots. 1897. Transferred to the 'N.S.M.O.' Dutch Flag. 1899. Broken up in Italy.
Sister ships: ACHILLES 1866. AJAX 1866, both by Scott and Co.

These three ships were ordered for a price of £156,000 and were the first to be built for the newly-registered Ocean Steam Ship Co. AGAMEMNON sailed on her maiden voyage to China on 19th April, 1866, followed by AJAX in June and ACHILLES in September. The route was Liverpool, Mauritius, Penang, Hong Kong, Shanghai. On the return voyage, tea was loaded at Foochow.
(Merseyside County Museums, Liverpool)

NESTOR, 1868.

1868 by Andrew Leslie and Co., Newcastle for Ocean Steam Ship Co. 1,869 G.R.T. 314′ x 33′. Compound steam-engines, 450 I.H.P. Single-screw. 9½ knots. 1894 February. Sold to Japanese owners. 1894 November. Destroyed by fire.

Slightly longer but with less beam than the previous three ships and possessing a smaller engine in the belief that the AGAMEMNON class was overpowered for the China run. A further ship, DIOMED, was constructed in 1868. She was similar to NESTOR but her specifications were slightly different: 291.5′ x 34.5′ and an engine of 630 I.H.P.
(Merseyside County Museums, Liverpool)

GLAUCUS, 1871

1871 by Andrew Leslie and Co., Newcastle for Ocean Steam Ship Co. 2,074 G.R.T. 322′ x 34′. Compound steam-engines, 700 I.H.P. Single-screw. 10 knots. 1891. Transferred to 'N.S.M.O.' Dutch flag. Same name. 1895. Sold to Japanese owners. Renamed JINTSU MARU. 1898. June 28th. Wrecked.

Sister ships: PATROCLUS 1872, DEUCALION 1872, ANTENOR 1872. All by Leslie and Co. The four were part of a nine-ship building programme that Holts undertook to combat the increased competition from other companies when the Suez Canal opened in 1869. The remaining ships were PRIAM 1870, SARPEDON and ULYSSES 1870–1, HECTOR and MENELAUS 1871.

(Merseyside County Museums, Liverpool)

ANCHISES, 1875.

1875 by Scott and Co., Greenock for Ocean Steam Ship Co. 2,021 G.R.T. 314′ x 35′. Compound steam-engines. 729 I.H.P. Single-screw, 9½ knots. 1891 September. Transferred to 'N.M.S.O.'. Dutch flag. Same name. 1896 July. Sold to Chinese owners. Wrecked near Rangoon.
Sister ships: STENTOR 1875, ORESTES 1875. Both by Scott and Co.

The three ships were the first to be designed with two funnels. This feature was never repeated on any of the ships designed and built specifically for the company.
(Merseyside County Museums, Liverpool)

TEUCER, 1877.

1877 by Scott and Co., Greenock for Ocean Steam Ship Co. 2,057 G.R.T. 317′ x 35′. Compound steam-engines. 788 I.H.P. Single-screw. 11 knots. 1885 May. Wrecked off Ushant, homeward bound with Sumatra tobacco.
Sister ship: ORESTES 1877, Scott and Co. This ship replaced the ORESTES of 1875 which was wrecked off Ceylon, 7th March, 1876.
(Merseyside County Museums, Liverpool)

5

LAERTES, 1879.

1879 by Scott and Co., Greenock for Ocean Steam Ship Co. 2,148 G.R.T. 321′ x 34′. Compound steam-engines. 777 I.H.P. Single-screw. 10 knots. 1894 November. Transferred to 'N.S.M.O.'. Dutch flag. Same name. 1901. Reverted to Ocean Steam Ship Co. 1903. Sold to Chinese. 1917 December. Sunk by collision in Malacca Strait.
Sister ships: CYCLOPS 1880, BELLEROPHON 1880. Both by Scott and Co. TELEMACHUS 1880. JASON 1880. Both by Andrew Leslie and Co.
(Merseyside County Museums, Liverpool)

PRIAM, 1890.

1890 by Scott and Co., Greenock for Ocean Steam Ship Co. 2,846 G.R.T. 336′ x 39′. Compound steam engines. 1,560 I.H.P. Single-screw. 11 knots. 1899. Transferred to 'N.S.M.O.'. Dutch flag. Same name. 1903. Sold to Japanese. Renamed SHINGU MARU. 1944 May 3rd. Bombed and sunk by U.S. aircraft, S.W. of Formosa. Sister ships: MYRMIDON 1890. TEUCER 1890. POLYPHEMUS 1890. All by Scott and Co. (Merseyside County Museums, Liverpool)

IXION, 1892.

1892 by Scott and Co., Greenock for Ocean Steam Ship Co. 3,572 G.R.T. 355′ x 43′. Triple-expansion engines. 2,286 I.H.P. Single-screw. 11 knots. 1902 September. Transferred to 'N.S.M.O.'. Dutch flag. Same name. 1911. October 2nd. Whilst on passage from Java to Amsterdam, caught fire off Engano, S.W. Sumatra. Abandoned and sank; the first loss for 21 years.

Sister ships: TANTALUS 1892, ULYSSES 1892, PYRRHUS 1892. All by Scott and Co.

Built of steel and with triple-expansion engines. The class was designed to carry heavier cargoes with no passenger accommodation provided.

(Merseyside County Museums, Liverpool)

ORESTES, 1894.

1894 by Scott and Co., Greenock for Ocean Steam Ship Co. 4,653 G.R.T. 392′ x 47′. Triple-expansion engines. 2,600 I.H.P. Single-screw. 11½ knots. 1925. Sold to Italian ship-breakers.

The first Holt ship to sail on the direct Glasgow to Australia service via the Cape in 1901.

Sister ships: DARDANUS 1894. DIOMED 1895. MENELAUS 1895. All by Scott and Co. HECTOR 1895. SARPEDON 1895. Both by Workman, Clark.

(Photograph by J. Clarkson)

PATROCLUS, 1896.

1896 by Workman, Clark, Belfast, for Ocean Steam Ship Co. 5,312 G.R.T. 422′ x 49.4′. Triple-expansion engines. 4,000 I.H.P. Single-screw. 12½ knots. 1907 September. Aground on Portland Bill. 1914. Transferred to 'N.S.M.O.'. Dutch flag. Same name. 1922. Renamed PALAMED. 1924. Renamed AUSTRALIA when sold to Italian owners. 1929. Scrapped at Genoa.
Sister ships: PROMETHEUS 1896, GLAUCUS 1896. Both by Scott and Co. ANTENOR 1896. Workman, Clark.
(Photographer unknown – Author's collection)

PATROCLUS, 1896.
A view of PATROCLUS aground on Portaland Bill, taken from a postcard posted in Weymouth 16th September 1907. The ship was homeward bound from Australia when she went ashore in fog. Refloated after 10 days with the aid of an especially fitted salvage ship from Liverpool under the personal supervision of the Company's naval architect, Henry Wortley.
(Publisher unknown – Author's collection)

IDOMENEUS, 1899.

1899 by Scott and Co., Greenock for Ocean Steam Ship Co. 6,497 G.R.T. 442' x 53'. Triple-expansion engines. 4,000 I.H.P. Single-screw. 12½ knots. 1922. Transferred to 'N.S.M.O.'. Dutch flag. 1925. Sold to Ditta L Pittaluga Vapori, Genoa. Renamed AURANIA. 1933. Scrapped at Spezia.
Sister ships: CALCHAS 1899, MACHAON 1899, ALCINOUS 1900. All by Scott and Co. STENTOR 1899. Workman, Clark.

This class was designed with minimum sheer to gain longitudinal strength, and also incorporated a prefabricated propeller-arch and a slightly buoyant hollow rudder to improve steering.
(Photograph by A. Duncan)

AJAX, 1900.
1900 by Scott and Co., Greenock for Ocean Steam Ship Co. 7,040 G.R.T. 442′ x 53′. Triple-expansion engines. 4,000 I.H.P. Single-screw. 12½ knots. 1915 October 10th. Shelled by a submarine in the Mediterranean, and saved when a British warship approached, which enabled her to escape. 1930. Sold for breaking up in Japan.
Sister ships: AGAMEMNON 1900, ACHILLES 1900. DEUCALION 1900. All by Scott and Co.
Similar design to preceding class and the first ships to exceed 7,000 G.R.T.
(Photograph by J. Clarkson)

The Suez Canal.

An early postcard of a Blue Funneller of about this period (c. 1900) passing through the Suez Canal. Opened in 1869, it shortened the distance to the Far East by some 3,000 miles, a saving of 10–12 days sailing time, an advantage against which even the fast clippers could not compete, but at the same time enabled other steam-ship companies with less efficient ships to make the voyage to China and beyond.
(Postcard – Author's collection)

JASON, 1902.
1902 by Workman, Clark, Belfast for Ocean Steam Ship Co. 7,450 G.R.T. 455′ x 54′. Triple-expansion engines. 4,000 I.H.P. Single-screw. 12½ knots. 1931 August. Sold for breaking up in Japan.
Sister ships: PELEUS 1901, TYDEUS 1901, TELEMACHUS 1902. All by Workman, Clark.

The last class to have the prefabricated propeller-arch, although the same arrangement was used in subsequent ships, but as a casting not a fabrication.
(Photograph by J. Clarkson)

PAK LING, 1902.

1895 by Workman, Clark, Belfast, for China Mutual Steam Navigation Co. 4,621 G.R.T. 410′ x 48′. Triple-expansion engines. 3,600 I.H.P. Single-screw. 12½ knots. 1902. Acquired by Alfred Holt. Same name. 1923. Broken up in Germany. Similar ship: KINTUK 1895. Workman, Clark. (Photographer unknown – Author's collection)

KEEMUN, 1902.

1902 by Workman, Clark, Belfast, for China Mutual Steam Navigation Co. 9,074 G.R.T. 481′ x 58′. Triple-expansion engines. 5,500 I.H.P. Twin-screw. 13 knots. 1902. Acquired by Alfred Holt whilst being built, following the purchase of China Mutual Steam Navigation Co. 1918 June 13th. Successfully fought off a U-boat in the Atlantic whilst being attacked by gunfire. 1933. Sold to Japan for scrapping.
Similar ships: NINGCHOW 1902. OANFA 1903. Both by Henderson of Glasgow.
(Photograph by J. Clarkson)

TELEMON, 1904.
1904 by Workman, Clark, Belfast for Ocean Steam Ship Co. 4,543 G.R.T. 383′ x 47′. Triple-expansion engines. 2,700 I.H.P. Single-screw. 11½ knots. 1933. Sold for scrapping at Port Glasgow.
Sister ships: PRIAM 1904, LAERTES 1904, by Hawthorn Leslie, Newcastle.
(Photograph by J. Clarkson)

MEMNON, 1906.
1906 by Scotts S.B. & E. Co., Glasgow for China Mutual Steam Navigation Co. 4,870 G.R.T. 392′ x 49′. Triple-expansion engines. 2,500 I.H.P. Single-screw. 11 knots. 1930. Scrapped at Kobe, Japan.
Sister ships: MYRMIDON 1905, POLYPHEMUS 1906, by Armstrong Whitworth, Newcastle.
ASTYANAX 1906, by Scott S.B. & E. Co.
(Photograph by J. Clarkson)

TEUCER, 1906.
'BELLEROPHON Class'.

1906 by Hawthorn Leslie, Newcastle for Ocean Steam Ship Co. 9,079 G.R.T. 485' x 58'. Triple-expansion engines. 5,700 I.H.P. Twin-screw. 12½ knots. 1915 December. Her speed enabled her to escape from a surfaced U-boat in the Mediterranean. 1944. Took part in the invasion of Sicily. 1948. Scrapped at Troon.
Sister ships: BELLEROPHON 1906, Workman, Clark. TITAN 1906, CYCLOPS 1906. Henderson, Glasgow. ANTILOCHUS 1906, Hawthorn Leslie.

Designed for the Pacific service and the transportation of long heavy logs, hence the 'goal-post' masts which enabled the derricks to lift the heavy loads as far outboard as possible. The first ships built for Ocean Steam Ship Co. with twin-screws.
(Photograph by J. Clarkson)

PROTESILAUS, 1910.

1910 by Hawthorn Leslie, Newcastle for China Mutual Steam Navigation Co. 9,547 G.R.T. 485' x 60'. Triple-expansion engines. 5,700 I.H.P. Twin-screw. 13½ knots. 1940 January 21st. Mined in the Bristol Channel close to the Welsh coast and abandoned. Although brought into port, found to be uneconomical to repair and broken up at Briton Ferry in 1942.

First of another three 'foot-ball' ships for the trans-Pacific service, though slightly smaller than the two following ships which came two years later: TALTHYBIUS 1912, IXION 1912, both by Scotts S.B. & E. Co.
(Photograph by J. Clarkson)

21

S.S. "Aeneas" Fishguard Harbour November 19th 1910

AENEAS, 1910.

During 1910–11 the Company had a class of three ships built for the Ocean Steam Ship Co., these being AENEAS 1910, ASCANIUS 1910 and ANCHISES 1911 all from Workman, Clark, Belfast. They were for a new six-weekly service to Australia via the Cape with refrigerated space for meat and fruit. Besides being the first ships to exceed 10,000 G.R.T. they also had the largest passenger accommodation of any ship built for the Company to date.

This card, believed to be the first used by the company for publicity purposes, shows AENEAS in Fishguard Harbour which was used for passenger embarkation from 1910 until 1913.

(Blue Funnel Line postcard – Tom Stanley collection)

The Blue Funnel Line issued several different cards for publicising the AENEAS Class. This example is by Norman Wilkinson, dated 1910 by the artist, advertising the service to South Africa and Australia.

The same design was used for a set of three postcards, each bearing the name of one of the ships in place of the advertisement. (Blue Funnel Line postcard – Author's collection)

The Blue Funnel Line,
Great Britain, South Africa and Australia.

AENEAS, 1910.
'AENEAS Class'

1910 by Workman, Clark, Belfast for Ocean Steam Ship Co. 10,048 G.R.T. 493' x 60.5'. Triple-expansion engines. 5,700 I.H.P. Twin-screw. 14 knots. Passengers: 150 Saloon. 1910 August 23rd – Launched; November 18th – Maiden voyage Glasgow/Liverpool to Brisbane. 1914. Troop transport. 1920 May 29th. First post-war voyage Glasgow to Brisbane. 1925. Transferred to Blue Funnel Line's Far East service. 1940 July 2nd. Bombed and sunk by German aircraft in the English Channel.

The postcard is from a set of three all bearing the same picture, but each with a different ship's name on the reverse.
(Blue Funnel Line postcard – Author's collection)

S 9198

BLUE FUNNEL LINE, S. S. "ASCANIUS."

ASCANIUS, 1910.
'AENEAS Class'

1910 by Workman, Clark, Belfast for Ocean Steam Ship Co. 10,048 G.R.T. 493′ x 60′. Triple-expansion engines, 5,700 I.H.P. Twin-screw. 14 knots. Passengers: 150 Saloon. 1910 December 30th. Maiden voyage Glasgow/Liverpool to Brisbane. Served as a troop-ship in both world wars. 1944 July 30th. Torpedoed in the English Channel. Repaired at Liverpool. 1945. Marseilles to Haifa service. 1949. Sold to Italy. Panama flag. Laid up. 1952. Broken up at La Spezia.

The postcard is one of a three-card series published by Kingsway Real Photo.
(Kingsway Real Photo – Author's collection)

T.S.S. ASCANIUS

A Blue Funnel postcard using a watercolour painted by Dawson It is possible that the ship represents the AENEAS class of 1910.
(Blue Funnel Line Postcard – Tom Stanley Collection)

ASCANIUS, 1910.
'AENEAS Class'

An interesting and unusual Blue Funnel postcard of Ascanius as a troop-ship during World War 1. The caption reads: 'Departure of South Australian Infantry of the First Australian Expeditionary Force from Outer Harbour S.A. 20th October 1914'.
(Blue Funnel Line postcard – S. Benz collection)

"S.S. ASCANIUS."
Departure of South Australian Infantry of the First Australian Expeditionary Force from Outer Harbour S.A. 20 October 1914

T.S.S. ULYSSES.   14,626 Tons.

NESTOR AND ULYSSES, 1913.

The popularity of the three preceding ships on the Australia run induced the Company to order two larger ships and to commence a monthly service. They exceeded their three running partners by almost 4,500 G.R.T. and had a deadweight of 16,000 tons.

The illustration is from a coloured postcard issued by the Company for both ships, the only difference being the name with attendant tonnage. There are, however, at least three different printings, each printing having a difference in colouring and an alteration to the tonnages.
(Blue Funnel Line postcard – Author's collection)

NESTOR, 1913.

1913 by Workman, Clark, Belfast for Ocean Steam Ship Co. 14,628 G.R.T. 563′ x 68′. Triple-expansion engines. 7,750 I.H.P. Twin-screw. 14 knots. Passengers: 240 Saloon. 1912 December 7th. Launched. 1913 May. Maiden voyage Glasgow/ Liverpool to Brisbane. 1915. Troop transport. 1920 April. First post-war voyage Glasgow to Brisbane. 1950 July. Broken up at Faslane by British Iron and Steel Corporation.
Sister ship: ULYSSES 1913. Workman, Clark.
(Postcard publisher unknown – Author's collection)

ULYSSES, 1913.

1913 by Workman, Clark, Belfast for China Mutual Steam Navigation Co. 14,646 G.R.T. 563′ x 68′. Triple-expansion engines. 7,750 I.H.P. Twin-screw. 14 knots. Passengers: 240 Saloon. 1913 July 5th. Launched 1913 October. Glasgow/Liverpool to Brisbane service. 1915. Troop transport. 1927 September. First post-war voyage Glasgow to Brisbane. 1945 April 11th. Torpedoed by U-160, 45 miles south of Cape Hatteras. She quickly sank but no lives were lost. Sister ship: NESTOR 1913, Workman, Clark.
(Photograph by A. Duncan)

A Blue Funnel Line publicity postcard showing NESTOR/ULYSSES at Capetown.
(Author's collection)

The Company had a series of ships built between 1908 and 1917 falling into two classes. The first class at 6,700 G.R.T were; PERSEUS 1908, THESEUS 1908, NELEUS 1911, ATREUS 1911, RHESUS 1911, DEMODOCUS 1912, LAOMEDON 1912, EUMAEUS 1913, PHEMIUS 1913. Followed by a second class, slightly longer and at 7,400 G.R.T.; LYCAON 1913, HELENUS 1913, TROILUS 1914, TEIRESIAS 1914, AGAPENOR 1914, MENTOR 1914, PYRRHUS 1914, ELPENOR 1917, TROILUS 1917, DIOMED 1917.

After World War 1, there was a shortage of passenger accommodation to the Far East. Several Holt ships had temporary accommodation provided and this Company postcard by James S. Mann was possibly issued about this time to appeal to passengers carried in any of the above ships.
(Blue Funnel Line postcard – Author's collection)

DOCKING "RHESUS"

RHESUS, 1911.

1911 by Scotts S.B. & E. Co., Greenock for China Mutual Steam Navigation Co. 6,719 G.R.T. 443' x 53'. Triple-expansion engines. 4,600 I.H.P. Single-screw. 13 knots. 1917 July 14th. Torpedo attack by U-boat off S.W. Ireland; no damage or loss of life. 1939 September and October. Employed with other ships in the transportation of the British Expeditionary Force to French ports. 1950. Scrapped at Port Glasgow.

An example of the first, slightly smaller class of ship built between 1908 and 1913. (Postcard publisher unknown – Author's collection)

DIOMED, 1917.

1917. Scotts S.B. & E. Co., Greenock for Ocean Steam Ship Co. 7,523 G.R.T. 455′ x 56′. Single reduction geared turbines. 5,700 S.H.P. Single screw. 14½ knots. 1918. August 21st. Sunk by gunfire from U-boat, 195 miles E.S.E. of Nantucket.

Although a member of the second, larger class, she differed from her sister ships by being the first Holt ship to have turbines and was also possibly the fastest up to that time. Commencing service in 1917 and sunk the following year, it is reasonable to assume that she never appeared in Blue Funnel colours; being camouflaged as the photograph shows.
(Photograph by A. Duncan)

The temporary passenger accommodation, fitted to the ships of this class, was quite successful, so much so that it was retained as a permanent feature on the MENTOR 1914, PYRRHUS 1914, and TEIRESIAS 1914, which are collectively shown on this postcard. The postcard illustrated was posted in Liverpool on 4th August 1922 to an address in Switzerland.
(Blue Funnel Line postcard – Author's collection)

S.S. "Pyrrhus"   The Blue Funnel Line

PYRRHUS, 1914.

1914 by Workman, Clark, Belfast for Ocean Steam Ship Co. 7,417 G.R.T. 455′ x 56′. Triple-expansion engines. 5,500 I.H.P. Single-screw. 13½ knots. 1940 February 17th. Torpedoed and sunk by U-boat, about 100 miles N.W. of Cape Finisterre. Eight lives lost.

From the existence of this card it was probable that the Company also issued publicity cards for the individual ships, to make a set of three besides the collective one reproduced on the previous page.
(Blue Funnel Line postcard – Author's collection)

EURYLOCHUS, 1915.

Built 1912 as INDRAGHIRI FOR T.B. Royden & Co., (Indra Line) by London and Glasgow Co., Glasgow. 5,723 G.R.T. 431' x 54'. Triple-expansion engines. 2,500 I.H.P. Single-screw. 12 knots. 1915. Purchased by Alfred Holt and Co. Operated by China Mutual Steam Navigation Co. and renamed. 1918 July 22nd. Chased by a surfaced U-boat which opened fire but EURYLOCHUS escaped. 1941 January 29th. Sunk by raider KORMORAN off West Africa.

Ships of the Indra Line acquired by Holts in 1915 were: INDRASAMHA, INDRAWADI, INDRADEO, INDRAGHIRI, INDRAKUALA, INDRA, INVERCLYDE.

(Photograph by J. Clarkson)

KNIGHT COMPANION, 1917.

Built 1913 by Charles Connell, Glasgow for Greenshields, Cowie and Co. (Knight Line). 7,375 G.R.T. 470′ x 58′. Triple-expansion engines. 3,000 I.H.P. Single-screw. 11½ knots. 1917. Purchased by Alfred Holt and Co. Operated by Ocean Steam Ship Co. Same name. 1917 June 11th. Torpedoed in the Atlantic but managed to reach Liverpool under tow. 1933. Scrapped in Italy.

Somewhat larger than the vessels of the Indra Line, the three others of the Knight Line purchased by Alfred Holt and Co. in 1917 were: KNIGHT OF THE GARTER, KNIGHT OF THE THISTLE and KNIGHT TEMPLAR. They were not renamed.
(Photograph by J. Clarkson)

TYNDAREUS, 1916.

1916 by Scotts S.B. & E. Co., Greenock for Ocean Steam Ship Co. 11,346 G.R.T. 507′ x 63′. Triple-expansion engines. 6,000 I.H.P. Twin-screw. 14 knots. Steerage passengers. 1916. Completed and entered service as troop transport. 1917 February 6th. Mined off Cape Agulhas. The mines had been laid during January by the raider WOLF. The damaged ship was towed with some difficulty to Capetown. Entered trans-Pacific service after World War 1. 1940. Troop ship and supply vessel. 1950. Indonesia-Jeddah pilgrim service. 1960 September. Arrived Hong Kong for scrapping.
(Photograph by A. Duncan)

TYNDAREUS, 1916.
A Company postcard of TYNDAREUS as a pilgrim ship. She was converted for this trade in 1949, deck and dormitory accommodation being provided for 2,500 pilgrims. Between seasons she was laid up at Singapore.
(Blue Funnel Line postcard – Author's collection)

LAERTES, 1919.

1919 by Taikoo Dockyard, Hong Kong for Ocean Steam Ship Co. 5,825 G.R.T. 420′ x 52′. Triple-expansion engines. 3,500 I.H.P. Single-screw. 13 knots. 1922. Transferred to 'N.S.M.O.' Dutch flag. Same name. 1940 February 3rd. Struck a mine off the Royal Sovereign lightship, slight damage sustained. 1942 May 3rd. Torpedoed off Cape Canaveral, Florida. 18 lives lost.
Sister ship: AUTOLYCUS 1917. By Taikoo Dockyard.
(Photograph F.W. Hawks collection)

ACHILLES, 1920.

1920 by Scotts S.B. & E. Co., Greenock for Ocean Steam Ship Co. 11,404 G.R.T. 507′ x 63′. Steam-turbines. 7,000 S.H.P. Twin-screw. 14 knots. 1940 August. Requisitioned by the Royal Navy, renamed H.M.S. BLENHEIM, and used as a destroyer depot ship. 1948 Scrapped at Barrow.

Another 'foot-ball' ship, she was, together with her sister-ship, placed on the trans-Pacific service. Both ships purchased by the Admiralty in 1940. Sister ship: PHILOCTETES 1922. By Scotts S.B. & E. Co.

(Photograph by J. Clarkson)

HECUBA, 1921.

1901 by Bremer Vulcan, Vegesack, Germany. 7,747 G.R.T. 431′ x 54′. Quadruple-expansion engines. 3,200 I.H.P. Twin-screw. 12½ knots.

Built in 1901 as BRANDENBURG for Norddeutscher Lloyd, this ship had been taken over by the British Government as reparation for war losses. It was purchased in 1921 by Holts who used her as a troop-ship mainly between Southampton and India. 1924. Sold for breaking up in Italy.
(Postcard Publisher unknown – Author's collection)

HECUBA, 1921.
A photograph dated June 1924 showing HECUBA laid up on the River Fal from where she was sold for scrap.
(Photographer unknown – Author's collection)

TROILUS, 1921.

1921 by Scotts S.B. & E. Co., Greenock for China Mutual Steam Navigation Co. 7,421 G.R.T. 459′ x 56′. Double-reduction geared turbines. 6,000 S.H.P. Single-screw. 14½ knots. 1942 June. Malta Convoy. 1944 September 1st. Torpedoed by submarine in the Indian Ocean, homeward bound from Colombo. During the period 1920–23 eleven similar ships were built. The first, MACHAON 1920 had triple-expansion engines, followed by EUMAEUS 1921. PHEMIUS 1921. TROILUS 1921. GLAUCUS 1921. AUTOLYCUS 1922. MERIONES 1922. AUTOMEDON 1922. RHEXENOR 1922 all with double-reduction geared turbines. The latter four were also wider in the beam by 2 feet.

(Photograph by A. Duncan)

ADRASTUS, 1923.

1923 by Scotts S.B. & E. Co., Greenock for Ocean Steam Ship Co. 7,905 G.R.T. 460′ x 58′. Single-reduction geared turbines. 6,000 S.H.P. Single-screw. 14½ knots. 1951 July. Renamed EURYADES. 1954 August. Sold for scrappng at Faslane.

This ship together with DARDANUS 1923 were the last two ships of this series of eleven. They differed from the others by having single-reduction geared turbines. Shown with double-tiered lifeboats amidships whilst in service as a pilgrim ship.
(Photograph by J. Clarkson)

PERSEUS, 1923.

1923 by Caledon S.B. & E. Co., Dundee for China Mutual Steam Navigation Co. 10,286 G.R.T. 491′ x 62′. Double-reduction geared turbines. 6,500 S.H.P. Twin-screw. 14½ knots.

One of four similar 'foot-ball' ships introduced for the Liverpool-Far East service. 1940 May. Left Holland with part of the Dutch gold reserves. 1944 January 16th. Torpedoed off the east coast of India by the Japanese submarine I-165. Entire crew rescued the same day by Naval ships. No lives lost.

Sister ships: CALCHAS 1921, DIOMED 1922. Both by Workman, Clark. MENELAUS 1923. By Caledon S.B. & E. Co.

(Photograph by J. Clarkson)

SARPEDON Class 1923–1925.

Several different designs of postcard were issued by Blue Funnel for the four ships of the SARPEDON Class built in 1923–25. This splendid watercolour was painted by the artist, Walter Thomas. Ordered for the Liverpool-Far East service, the four 15-knot turbine steamers accommodated 150 first-class passengers and became very popular ships over the years. This card, overprinted on the reverse for the PATROCLUS was posted in Singapore on 1st August 1939 to an address in Colombo, Ceylon.

(Blue Funnel postcard – Author's collection)

Painted by a different artist and showing another publicity card illustrating one of the four ships of the SARPEDON Class built in 1923–25. The ships were as follows:

SARPEDON 1923 by Cammell Laird, Birkenhead, for Ocean Steam Ship Co. 1953 June. Sold for breaking up at Newport, Monmouthshire. PATROCLUS 1923 by Scotts S.B. & E. Co., Greenock for China Mutual Steam Navigation Co. 1940 November 4th. Torpedoed and sunk by U-99 off the west coast of Ireland whilst serving as an armed merchant cruiser. HECTOR 1924 by Scotts S.B. & E. Co., Greenock for Ocean Steam Ship Co. 1942 April 5th. Bombed and sunk in Colombo harbour whilst serving as an armed merchant cruiser. ANTENOR 1925 by Palmers, Newcastle for China Mutual Steam Navigation Co. 1953 July. Sold for breaking up at Blyth.
(Blue Funnel postcard – Author's collection)

A postcard showing the launch of either SARPEDON or PATROCLUS in 1923. Both ships were fitted with derrick posts at the after end of the boat deck. Neither HECTOR or ANTENOR were fitted with them. Further particulars of the class were: SARPEDON 11,321 G.R.T. 499' x 62'. Coal-fired.   PATROCLUS 11,314 G.R.T. 499' x 62'. Coal-fired. HECTOR 11,198 G.R.T. 499' x 62'. Oil-fired.   ANTENOR 11,174 G.R.T. 499' x 62'. Oil-fired. All four ships were twin-screw, single-geared turbines, 7,500 S.H.P. 15 knots. (Postcard publisher unknown – Author's collection)

The Blue Funnel Line issued several publicity postcards for advertising its passenger services. This example, issued in 1927, was the only time a young lady was seen on any company postcards. Another printing of the same scene displays details of the passenger and cargo services to Egypt, Ceylon, Straits, China, Japan, South Africa, Australia, Java and America. Sam Brown was the artist responsible for the ship and possibly the general view, with another artist painting the young lady.

(Blue Funnel postcard – Author's collection)

MEDON, 1923.

1923 by Palmers, Newcastle for Ocean Steam Ship Co. 5,444 G.R.T. 407' x 52'. Motor-ship. 2,500 B.H.P. Single-screw. 11 knots. 1942 August 10th. Torpedoed in mid-Atlantic by the Italian submarine GIULIANI.

The first Blue Funnel motor-ship. TANTALUS 7,775 G.R.T., a twin-screw motor-ship and Caledon built, followed later in the same year.

(Photograph by J. Clarkson)

ASPHALION, 1924.

1924 by Scotts S.B. & E. Co., Greenock for China Mutual Steam Navigation Co. 6,274 G.R.T. 432' x 55'. Double-reduction geared turbines. 4,350 S.H.P. Single-screw. 13 knots. 12 passengers. 1944 February 11th. Struck by two torpedoes from a Japanese submarine whilst in convoy off Vizagapatam. Though badly damaged ASPHALION was towed to safety. Shortly after the incident, the submarine was destroyed. 1959. Scrapped at Hong Kong.
Sister ships: MELAMPUS 1924. Palmers, Newcastle. POLYDORUS 1925. Both by Scotts S.B. & E. Co. Following the delivery of this class, there were no more turbine-driven ships built for Holts for the next 24 years.
(Photograph F. W. Hawks' collection)

PROMETHEUS, 1925.

1925 by Scotts S.B. & E. Co., Greenock for Ocean Steam Ship Co. 6,094 G.R.T. 425′ x 56′. Motor-ship. 3,900 B.H.P. Twin-screw. 13½ knots. 1942 February 26th. Bombed but without serious damage, 270 miles S.S.W. of Rockall. 1957 November. Sold and renamed JANUS, Liberian flag. 1959. Sold for scrapping in Italy after being damaged by fire at Alicante on 12th October 1958.
Sister ships: EURYMEDON 1924. PEISANDER 1925. Both by Caledon.
(Photograph by F. R. Sherlock)

PHRONTIS, 1925.

1925 by Caledon S.B. & E. Co., Dundee for 'N.S.M.O.' 6,181 G.R.T. 425' x 55'. Motor-ship. 3,700 B.H.P. Twin-screw. 13½ knots. 12 Passengers. 1958. Sold and renamed RYAD. Saudi-Arabian flag. 1958 August. Arrived at Hong Kong for scrap.

Sister ships: ALCINOUS 1925. STENTOR 1926. Both by Scotts S.B. & E. Co.

These ships were a complete departure from the usual Holt design, with PHRONTIS and ALCINOUS being operated by the 'N.S.M.O'. Dutch flag, whilst the STENTOR was for the China Mutual Steam Navigation Co.

(Photograph by Fotoship, Penarth)

ORESTES, 1926.

1926 by Workman, Clark, Belfast for Ocean Steam Ship Co. 7,765 G.R.T. 458′ x 58′. Motor-ship. 6,000 B.H.P. Twin-screw. 14½ knots. 12 Passengers. 1942 April 2nd. Bombed by Japanese aircraft off Madras, no damage. 1942 June 9th. Shelled by Japanese submarine off Sydney, slight damage. 1944 July. Took part in the assault on Sicily. 1963 August. Arrived at Mihara, Japan, for scrap.
Sister ship: IDOMENEUS 1926. Workman, Clark.
(Photograph by T. Rayner)

EURYBATES, 1928.

1928 by Scotts S.B. & E. Co., Greenock for Ocean Steam Ship Co. 6,276 G.R.T. 432′ x 55′. Scott-Still engines. 5,000 B.H.P. Twin-screw. 13½ knots. This ship had been fitted with a Scott-Still engine after successfully being tried in the DOLIUS 1924. The principle was that heat from the diesels cooling sytem was used to produce steam for an engine working in combination with the diesel. 1951. Steam-engine removed to become motor-ship only. 1958. Sold for scrap at Ghent.
(Photograph by A. Duncan)

THE BLUE FUNNEL LINE   M.V. " AGAMEMNON "

AGAMEMNON, 1929.

A Blue Funnel Line postcard illustrating the principal ship of the five ships known as the AGAMEMNON class. Introduced for the Liverpool-Far East service and also round-the-world. 1929 by Workman, Clark, Belfast for Ocean Steam Ship Co. 7,829 G.R.T. 460′ x 59′. Motor-ship. 8,600 B.H.P. Twin-screw. 16 knots. 1940. Requisitioned by the Royal Navy as H.M.S. AGAMEMNON and serving as a mine-layer. 1944 November. Sailed to Vancouver to be converted to an 'Amenity ship' for the Pacific Fleet Train. 1946. Returned to commercial service. 1963. Broken up at Hong Kong. Sister ships: MENESTHEUS 1929. MEMNON 1931. Both by Caledon. DEUCALION 1930. By Hawthorn Leslie. AJAX 1931. By Scotts S.B. & E. Co.
(Blue Funnel Line postcard – Author's collection)

CLYTONEUS, 1930.

1930 by Scotts S.B. & E. Co., Greenock for Ocean Steam Ship Co. 6,278 G.R.T. 433′ x 56′. Motor-ship. 5,500 B.H.P. Twin-screw. 14½ knots. 1941 January 8th. Bombed by Fokker-Wulf Kondor off N.W. Ireland. Sank some hours later. Sister ships: MARON 1930. By Caledon. MYRMIDON 1930. POLYPHEMUS 1930. Both by Scotts S.B. & E. Co.

These four ships, all completed in 1930, were the last vessels to be added to the deep-sea fleet before World War 2. None of the ships saw the end of the war; all were lost during 1941–42.

(Photograph by A. Duncan)

GORGON, 1933.

1933 by Caledon S.B. & E. Co., Dundee for Ocean Steam Ship Co. 3,678 G.R.T. 320′ x 51′. Motor-ship. 4,000 B.H.P. Single-screw. 14 knots. 1943 April 14th. Bombed and damaged by Japanese aircraft at Milne Bay. Towed to Brisbane for permanent repairs. 1964. Sold for scrap at Hong Kong.
Similar ship: CHARON 1936. By Caledon.

These two near-sisters, together with the CENTAUR 1924, were used on the Australia-Singapore service.
(Blue Funnel Line postcard – Author's collection)

A Blue Funnel postcard by Laurence Dunn representing the GORGON and CHARON on the Singapore-Australia service. The CENTAUR 1924 had been torpedoed and sunk on 14th May, 1943 by a Japanese submarine while serving as a hospital ship.
(Blue Funnel Line postcard - Author's collection)

DEUCALION, 1970.

1939 by Caledon, S.B. & E. Co., Dundee for Glen Line. 9,919 G.R.T. 483′ x 66.5′. Motor-ship. 12,000 B.H.P. Twin-screw. 17½ knots. 12 passengers.

This ship was one of a class of eight ordered by Glen Line in 1936, after the company's acquisition by Blue Funnel, and delivered as GLENGYLE in 1939. Taken over by the Royal Navy within a few months of the outbreak of the war for conversion into a landing-ship for infantry. Returned to Glen Line after the war retaining the same name.
1970. Transferred to the Blue Funnel Line (Ocean Steam Ship Co.) renamed DEUCALION.
1971. Sold for scrap to Taiwan ship-breakers.
(Photograph by J. Clarkson)

STENTOR, 1946.

1946 by Caledon, S.B. & E., Co., Dundee for Ocean Steam Ship Co. 10,203 G.R.T. 496′ o.a. x 64′. Motor-ship. 6,800 B.H.P. Single-screw. 15½ knots. 12 passengers. 1958 November. Transferred to Glen Line, renamed GLENSHIEL. 1963. Reverted to Blue Funnel (China Mutual Steam Navigation Co.) renamed STENTOR. 1975. Sold to Taiwan ship-breakers for scrap. Renamed TENTO for the delivery voyage from Singapore.

Sister ship: RHEXENOR 1945. By Caledon.

Both vessels were to be used for the transportation of heavy equipment to the Far Eastern war zone. Purchased on the stocks and modified for the Australian trade.

(Photographer unknown – Author's collection)

MEDON, 1946.
1942 by Harland and Wolff, Belfast for Ocean Steam Ship Co. 7,362 G.R.T. 448′ o.a. x 57′. Motor-ship. 2,600 B.H.P. Single-screw. 12 knots. 1942. Built as EMPIRE SPLENDOUR for Ministry of War Transport. 1946. Purchased by Blue Funnel (Ocean Steam Ship Co.) renamed MEDON. 1963. Sold to Liberian owners. Renamed TINA. 1970. Broken up. Chinese mainland ship-breakers.
(Photographer unknown – Author's collection)

TALTHYBIUS, 1947.

1943 by Bethlehem Fairfield, Baltimore, Maryland, U.S.A. 7,317 G.R.T. 442'o.a. x 57'. Triple-expansion engines. 2,500 I.H.P. Single-screw. 11 knots. 1943. Launched as PETER COOPER. Transferred to Britain under 'Lease Lend' and renamed SAMARKAND. 1947. Purchased for Ocean Steam Ship Co. renamed TALTHYBIUS. 1954. Transferred to Glen Line as GLENIFFER. 1950. Sold to Liberian owners. Renamed DOVE. 1955. Renamed PATRAIC SKY. 1971. Scrapped at Split, Yugoslavia.

Some war-losses were replaced with the acqustion of six 'Victory' ships and eight 'Liberties'; the TALTHYBIUS being one of the latter. They all underwent considerable modifications to make them suitable for Blue Funnel Line service.
(Photographer unknown – Author's collection)

TALTHYBIUS, 1960.

1945 by Permanente Metals Corporation, Richmond, California, U.S.A. 7,672 G.R.T. 455′o.a. x 62′ Steam turbines. 6,000 S.H.P. Single-screw. 16 knots. 1945 Launched as SALINA VICTORY. 1946. Acquired by Blue Funnel Line. Renamed POLYDORUS and operated by 'N.S.M.O.'. Dutch flag. 1960. Transferred to Blue Funnel (Ocean Steam Ship Co.), Renamed TALTHYBIUS. 1971. Transferred to Elder Dempster Lines Ltd. for two voyages to West Africa. 1972. Broken up at Kaohsiung, Taiwan.

This TALTHYBIUS, an ex-Victory ship, was in the fleet at a different time to the ship shown on the page 65. (Photograph by Airfoto, Malacca.)

'A' or ANCHISES CLASS

A Blue Funnel postcard by Walter Thomas representing the numerous ships of the 'A' Class. Twenty-seven vessels were built between 1947 and 1958 as part of the company's post-war shipbuilding programme. All of the ships had accommodation for twelve passengers. The 'A' Class was sub-divided into 6 groups, classified by a mark number and each group differing slightly from the others.

'A' Class Mk. 1 – six ships (1947–1948)   'A' Class Mk. 4 – four ships (1953–1954)
'A' Class Mk. 2 – six ships (1948–1949)   'A' Class Mk. 5 – three ships (1955–1956)
'A' Class Mk. 3 – five ships (1951–1953)   'A'Class Mk. 6 – three ships (1956–1958)
(Blue Funnel Line postcard – Author's collection)

AENEAS, 1947.
Class 'A' Mk. 1

1947 by Caledon S.B. & E. Co., Dundee for Ocean Steam Ship Co. 7,641 G.R.T. 487′ o.a. x 62′. Motor-ship. 6,800 B.H.P. Single-screw. 15½ knots. 1972. Broken up at Kaohsiung, Taiwan. Class 'A' Mk. 1 ships: CALCHAS 1947. By Harland & Wolff. ANCHISES 1947, ACHILLES 1948. Both by Caledon. AGAPENOR 1947. ASTYANAX 1948. Both by Scotts.
(Photograph publisher unknown – Author's collection)

CALCHAS, 1947.
Class 'A' Mk. 1.

A photograph of CALCHAS on fire at Port Kelang, Malaysia in July 1973. Found to be damaged beyond economical repair, she was sold for scrap, arriving at Kaohsiung, Taiwan the following October.
1947 by Harland and Wolff, Belfast for Ocean Steam Ship Co. 7,639 G.R.T. 487′o.a. x 62′. Motor-ship. 6,800 B.H.P. Single-screw, 15½ knots. The first ship of the 'A' class to be delivered. 1957. Transferred to Glen Line, renamed GLENFINLAS. 1962. Returned to Blue Funnel Line, renamed CALCHAS.
(Photograph by Airfoto, Malacca)

CYCLOPS, 1948.
Class 'A' Mk. 2.

1948 by Scotts S.B. & E. Co., Greenock for Ocean Steam Ship Co. 7,709 G.R.T. 487′o.a. x 62′. Motor-ship, 7,300 B.H.P. Single-screw 15½ knots. 1975 July. Renamed AUTOMEDON. 1975 December. Transferred to Elder Dempster Lines, with buff funnel. 1977. Sold for breaking up by W. H. Arnott Young and Co. Ltd. Class 'A' Mk. 2 ships: CLYTONEUS 1948. By Caledon. ANTILOCHUS 1949. By Harland & Wolff. AUTOLYCHUS 1949. AUTOMEDON 1949. LAERTES 1949. All by Vickers Armstrong. The Mk. 2 ships were specially fitted out for the carriage of 1,000 or more pilgrims from the Far East to Jeddah.
(Photograph by Fotoship, Penarth)

CYCLOPS, 1948.
Class 'A' Mk. 2

When radar was first installed by the company, placing the radar scanner on the bridge-top, it was soon discovered that the proximity of the funnel caused an arc of radar blindness. To remedy this problem, the scanner was re-sited on the funnel, as the photograph shows.
(Photograph by J. Clarkson)

A Company postcard, based on a watercolour by Walter Thomas, showing one of the 'A' Class Mk. 2 ships on the pilgrim service.
(Blue Funnel Line postcard – Author's collection)

ATREUS, 1951.
Class 'A' Mk. 3.

1951 by Vickers Armstrong, Newcastle for China Mutual Steam Navigation Co. 7,800 G.R.T. 487′ o.a. x 62′. Motor-ship. 7,000 B.H.P. Single-screw. 16 knots. 1977. Sold to Sherwood Shipping Co., Monrovia. Renamed UNITED VALIANT. 1979. Broken up at Kaohsiung, Taiwan, arriving on 23rd February 1979. Class 'A' Mk. 3 ships: ASCANIUS 1950. By Harland & Wolff. BELLEROPHON 1950. By Caledon. ALCINOUS 1952, LAOMEDON 1953. Both by Vickers Armstrong.
(Photograph by Fotoship, Penarth)

LAOMEDON, 1953.
Class 'A' Mk. 3.

1953 by Vickers Armstrong, Newcastle for China Mutual Steam Navigation Co. 7,864 G.R.T. 487′ o.a. x 62′. Motor-ship. 7,000 B.H.P. Single-screw. 16 knots. 1977 March. Sold to Regent Navigation Corporation, Panama. Renamed ASPASIA. 1978. To Pakistan for breaking up at Gadani Beach.

As an alternative to siting the radar-scanner on the funnel to overcome radar blindness, other ships were fitted with a mast which enabled the scanner to be positioned clear of the funnel.
(Photograph by K. Byass)

ADRASTUS, 1953.
Class 'A' Mk. 4.

1953 by Vickers Armstrong, Newcastle for Ocean Steam Ship Co. 7,859 G.R.T. 487′o.a. x 62′. Motor-ship. 7,000 B.H.P. Single-screw. 16 knots. 1960. Transferred to 'N.S.M.O.' Same name. 1975. Reverted to Blue Funnel. 1978. Sold to Rhodeswell Shipping Co., Cyprus. Renamed ANASSA. 1981. Sold for breaking up at Gadani Beach, Pakistan. Class 'A' Mk. 4 ships: EUMAEUS 1953. By Caledon. LYCAON 1954. By Vickers Armstrong. ELPENOR 1954. By Harland & Wolff.
(Photographer unknown – Author's collection)

DYMAS, 1949.

1922 by Harland & Wolff, Glasgow for Glen Line. 9,461 G.R.T. 502′o.a. x 62′. Motor-ship, 5,000 B.H.P. Twin-screw. 13 knots. 12 passengers. 1922. Delivered to the Glen Line as GLENBEG. 1949. Transferred to Blue Funnel (Ocean Steam Ship Co.) and renamed DYMAS. 1954. Sold for breaking up at Dalmuir, arriving April 8th 1954.
Similar ships: DEUCALION (ex-GLENOGLE) 1920. DARDANUS (ex-GLENAPP) 1920. DOLIUS (ex-GLENSTRAE) 1922. All by Harland & Wolff, Glasgow.

These ships were displaced from the Glen Line as the new GLENEARN class were reconditioned after war service. These four older ships were transferred to replace Blue Funnel war losses.
(Photograph by J. Clarkson)

TEIRESIAS, 1950.

1950 by J. L. Thompson, Sunderland for Silver Line. 8,910 G.R.T. 474′o.a. x 62′. Steam-turbines. 6,800 S.H.P. Single-screw. 16 knots. Launched as SILVERELM but purchased by Blue Funnel during construction and completed for 'N.S.M.O.' Amsterdam, as TEIRESIAS. 1960. Transferred to Ocean Steam Ship Co. British flag and renamed TELE-MACHAS. 1971. Sold to Anax Shipping Co. Ltd. Cyprus and renamed AEGIS COURAGE. 1973. Broken up by Chinese ship-breakers.
Similar ships: ULYSSES (ex-SILVERHOLLY) 1949. TEUCER (ex-SILVERLAUREL) 1950. Both by J. L. Thompson. Purchased as new ships from Silver Line as replacements for war losses.
(N.S.M.O. Postcard – Author's collection)

TELEMACHUS, 1960.
A photograph of TELEMACHUS (Ocean Steam Ship Co.) ex-TEIRESIAS of 1950.
(Photograph by Airfoto, Malacca)

'P' Class

A postcard by Walter Thomas showing one of the 'P' Class steam-turbine ships designed for the Far Eastern service with accommodation for 29 first-class passengers. The ships of the 'P' Class were: PELEUS 1949, PYRRHUS 1949. Both by Cammel Laird, Birkenhead. PATROCLUS 1950, PERSEUS 1950. Both by Vickers Armstrong, Newcastle. The passenger accommodation was removed in 1967 (PYRRHUS 1964/5 following a fire at Liverpool) together with one life-boat each side.
The card is from a set of four all with the same scene but having the particular ship's name on the reverse side.
(Blue Funnel Line postcard – Author's collection)

THE BLUE FUNNEL LINE

One of the 'P' Class illustrated on a fine publicity postcard by Walter Thomas. The same scene was used for a set of four, one for each ship, the name printed on the reverse.
(Blue Funnel Line postcard – Author's collection)

'P' Class

A Blue Funnel Line postcard using a photograph of a 'P' Class ship, the name of which is not shown, perhaps indicating that it was a single card used for the complete class.
(Blue Funnel Line postcard – David Cope collection)

PATROCLUS, 1950.
'P' Class.
1950 by Vickers Armstrong, Newcastle for China Mutual Steam Navigation Co. 10,109 G.R.T. 516'o.a. x 68'. Steam-turbines, 14,000 S.H.P. Single-screw. 18 knots. 1972. Renamed PHILOCTETES. 1972/3. She made one voyage under the name of PHILOCTETES, departing from Swansea 21st November, 1972 and arriving at Kaohsiung, Taiwan, 12th February 1973 for scrapping. The photograph shows the ship in her original form, with three life-boats positioned on each side amidships.
(Photographer unknown – Author's collection)

'H' Class.

A Walter Thomas watercolour postcard showing one of the 'H' Class steam-turbine ships for the Australian trade . Similar to the 'P' Class, though slightly longer and with more refrigerated space. They also carried 29 first-class passengers. Ships of the class were: HELENUS 1949. HECTOR 1950. IXION 1951. All by Harland & Wolff, Belfast. JASON 1950. By Swan, Hunter, Wallsend. The passenger accommodation was removed in 1964 and also one life-boat from each side. The postcard is again from a set of four, one for each ship, the name printed on the reverse.

(Blue Funnel Line postcard – Author's collection)

'H' Class.
One of the 'H' Class passing under Sydney Harbour Bridge illustrated on a publicity postcard; one of a set of four. Each postcard using the same design but having a different ship's name printed on the reverse.
(Blue Funnel Line postcard – Author's collection)

IXION, 1951.
'H' Class.
1951 by Harland & Wolff, Belfast for Ocean Steam Ship Co. 10,125 G.R.T. 523′ o.a. x 69′. Steam-turbines. 14,000 S.H.P.
Single-screw. 18 knots. 1972. Broken up at Barcelona, Spain, arriving 12th March, 1972.
(Blue Funnel Line postcard – Author's collection)

NELEUS, 1953.
'NELEUS Class'

1953 by Caledon S.B. & E. Co., Dundee for China Mutual Steam Navigation Co. 7,800 G.R.T. 490'o.a. x 64'. Steam-turbines. 8,000 S.H.P. Single-screw. 16 knots. 1971. Transferred to Glen Line, same name. 1971. Sold by Glen Line to Akamas Shipping Co. Ltd., Cyprus and renamed AEGIS FABLE. 1972. Sold to Alicacnossos Shipping Co. Ltd., Cyprus and renamed AEGIS TRUST. 1974. Broken up by Chinese shipbreakers.
Ships of the NELEUS Class: NESTOR 1952, THESEUS 1955. Both by Caledon. A class of refrigerated ships built for the Australian trade.
(Photograph by A. Duncan)

DEMODOCUS, 1955.
Class 'A' Mk. 5.

1955 by Vickers Armstrong, Newcastle for Ocean Steam Ship Co. 7,968 G.R.T. 492′o.a. x 62′. Motor-ship. 8,000 S.H.P.
Single-screw. 16½ knots. 1970. Transferred to Glen Line, renamed GLENROY. 1972. Reverted to Blue Funnel Line
(Ocean Steam Ship Co.) as DEMODOCUS. 1973. Sold to Nan Yang Shipping Co., Macao, renamed HUNGSIA.
1979. To Peoples Republic of China and renamed HONG QI 137. Class 'A' Mk. 5 ships: DIOMED 1956. By Caledon.
DOLIUS 1956. By Harland & Wolff.
(See cover illustration)

ANTENOR, 1956.
Class 'A' Mk. 6.
1956 by Vickers Armstrong, Newcastle for Ocean Steam Ship Co. 7,974 G.R.T. 492′ o.a. x 62′. Motor-ship. 8,000 B.H.P. Single-screw. 16½ knots. 1970 November. Transferred to Glen Line and renamed GLENLOCHY. 1972 June. Returned to Blue Funnel Line, renamed DYMAS. 1973 April. Sold to Nan Yang Shipping Co, Macao, renamed KAIYUN.
Class 'A' Mk. 6 ships: ACHILLES 1957, AJAX 1958. Both by Vickers Armstrong.
These three ships were the last of the 'A' class.
(Photographer unknown – Author's collection)

GUNUNG DJATI, 1958.

A Blue Funnel Line postcard by Walter Thomas. There are at least two of these cards, the only difference being the colour of the flag flying from the foremast on each card. 1936 by Blohm & Voss, Hamburg for Deutsche Ost-Afrika Linie, named PRETORIA. 18,036 G.R.T. 576′ o.a. x 82′. Steam-turbines. 14,200 S.H.P. Twin-screw. 16 knots. 1945. British war prize. Managed by Orient Line for Ministry of War Transport. Renamed EMPIRE DOON. Troopship. 1948–49. Reconditioned and renamed EMPIRE ORWELL. Troopship. 1958. Chartered to the Pan Islamic S.S. Co., Karachi for pilgrim service. 1958 November. Sold to Blue Funnel Line (Ocean Steam Ship Co.) and refitted for pilgrims. Renamed GUNUNG DJATI. 1962. Sold to Indonesian Government, same service. 1979. Indonesian naval accommodation ship, renamed KRI TANJUNG PANDAN.
(Blue Funnel Line postcard – Author's collection)

MENELAUS, 1957.
'M' Class.
1957 by Caledon S.B. & E. Co., Dundee for Ocean Steam Ship Co. 8,539 G.R.T. 495' o.a. x 65'. Motor-ship. 8,500 B.H.P. Single-screw. 16½ knots. 12 passengers. 1962. Passenger accommodation removed. 1972. Transferred to Elder Dempster Lines Ltd. and renamed MANO. 1977. Renamed OTI. 1978. Sold to Leon Rivera Lines Co. Ltd., Cyprus and renamed ELSTAR. 1979. Sold for breaking up at Busan, South Korea.
Class 'M' ships: MENESTHEUS 1958, MACHAON 1959, MARON 1960. All by Caledon. MEMNON 1959. MELAMPUS 1960. Both by Vickers Armstrong.
The 'M' Class superseded the 'A' Class. The ships were larger and capable of faster service speeds.
(Photograph by A. Duncan)

MACHAON, 1959.
'M' Class.
1959 by Caledon S.B. & E. Co., Dundee for Ocean Steam Ship Co. 8,530 G.R.T. 495′ o.a. x 65′. Motor-ship. 8,500 B.H.P.
Single-screw. 16½ knots. 12 passengers. 1962. Passenger accommodation removed. 1975. Transferred to 'N.S.M.O.'
Dutch flag. 1977 August. Transferred to Elder Dempster Lines Ltd. and renamed OBUASI. 1978 June. Sold to Lenake
Shipping Co., Cyprus, renamed ELSEA. 1978 July. Sold to Tartan Shipping Ltd., Greece, renamed MED ENDEAVOR.
1979. Broken up at Kaohsiung, Taiwan.
Notice the mast placed forward of the funnel, topped by the radar scanner.
(Photograph by Airfoto, Malacca)

CENTAUR, 1964.

1964 by John Brown, Clydebank, for Ocean Steam Ship Co. 8,262 G.R.T. 481′ o.a. x 66.3′. Motor-ship. 16,500 B.H.P. 18 knots. 190 passengers. Built for the service between Singapore and Western Australia. 1973. Managed by Straits S.S. Co., Blue Funnel Line colours. 1981. September 15th. Final departure from Fremantle. 1982. Chartered to St. Helena Shipping Co. (Curnow Shipping) to replace their own ship, ST. HELENA, which had been requistioned for the Falklands war. 1984 January. Returned to Straits S.S. Co. 1985. Sold to Shanghai Haixing S. Co., China and renamed HAI LONG. 1986. Renamed HAI DA, same owners.
The first Blue Funnel ship to have a tapered funnel.
(Blue Funnel Line postcard – Author's collection)

CENTAUR, 1964.

A watercolour postcard by Lawrence Dunn. The CENTAUR was designed for a three-weekly service from Freemantle and the Western Australian ports to Singapore, carrying sheep and cattle as well as passengers. Sadly, her service life was terminated by the introduction of specialised livestock ships and the public forsaking the ships in favour of air travel. These factors resulted in the service becoming uneconomic.

(Blue Funnel Line postcard – Author's collection)

CENTAUR, 1964.
The photograph taken in the Spring of 1983, shows CENTAUR displaying the funnel colours of Curnow Shipping whilst on charter to them during the Falklands war of 1982, although still registered in Singapore. An option to purchase was not taken up and CENTAUR returned to Straits S.S. Co. when Curnow's own vessel, ST. HELENA, was released from war service.
(Photograph by FotoFlite, Ashford)

BLUE FUNNEL

"PRIAM" CLASS
UNITED KINGDOM-JAPAN

'PRIAM' Class.

A Blue Funnel Line publicity card painted by Laurence Dunn for the 'PRIAM' Class. Eight ships were built for the express cargo-liner service between the U.K./Europe and the Far East; four for Blue Funnel and four for the Glen Line. The Blue Funnel Line ships were: PRIAM 1966. PEISANDER 1967. PROTESILAUS 1967. PROMETHEUS 1967. All by Vickers Armstrong, Newcastle. The Glen Line Ships, which eventually came to Blue Funnel, were as follows: 1972 PEMBROKE-SHIRE became PHRONTIS and GLENFINLAS became PHEMIUS. 1973 GLENALMOND became PATROCLUS and RADNORSHIRE became PERSEUS.

(Blue Funnel Line postcard – Author's collection)

PRIAM, 1966.
'PRIAM' Class.

1966 by Vickers Armstrong, Newcastle for Ocean Steam Ship Co. 12,094 G.R.T. 563.8′ o.a. x 78′. Motor-ship. 22,500 B.H.P. Single-screw, 21 knots. 1978. Sold to C.Y. Tung enbloc with PEISANDER, PROMETHEUS, PROTESILAUS and renamed ORIENTAL CHAMPION. Panamanian flag. 1985 October 10th. Struck by Iraqi missile about sixty miles north of Bahrain and arrived there under tow on 20th October. 1985 December 11th. Left Bahrain under tow for Taiwan ship-breakers, arriving Kaohsiung 13th February 1986.
(Blue Funnel Line postcard – Author's collection)

PROTESILAUS, 1967.
'PRIAM' Class.

1967 by Vickers Armstrong, Newcastle for China Mutual Steam Navigation Co. 12,094 G.R.T. 564′ o.a. x 78′. Motor-ship. 22,500 B.H.P. Single-screw. 21 knots. 1978. Sold to C.Y. Tung, renamed ORIENTAL IMPORTER. After this date until 1984, the ship had various owners. 1985 June 1st. Sustained considerable damage aft when struck by two rockets, believed to be Iranian, and caught fire whilst on a voyage from Damman to Kuwait. The fire was extinguished the following day but the ship was declared a total loss. 1985 July 25th. Left Bahrain for scrapping at Kaohsiung, Taiwan.
(Photographer unknown – Author's collection)

ORIENTAL IMPORTER, 1978.
A view of ORIENTAL IMPORTER, ex-PROTESILAUS of the 1967 'PRIAM' Class. The only photograph in the book
which is not of a Blue Funnel ship, but showing the modifications she underwent for conversion to a full container ship.
(Photograph by Airfoto, Malacca)

ACKNOWLEDGEMENTS

The author is indebted to the following people and sources for their best wishes and generous permission to use their photographs and postcards.

Mr. P. E. Hughes, Deputy Company Secretary, Ocean Transport and Trading Ltd., Liverpool.
Merseyside County Museums, Liverpool.

S. Benz
K. Byass
J. Clarkson
D. Cope
A. Duncan
Airfoto, Malacca, Malaysia
FotoFlite, Ashford, Kent
Fotoship, Penarth, S. Glamorgan.

F. W. Hawks
T. Rayner
F. R. Sherlock
T. Stanley

Every effort has been made to identify copyright material.

My thanks to Steve Benz, from whom the idea for this book originally came, for his unfailing enthusiasm, guidance and editorial.

BIBLIOGRAPHY

Great Passenger Ships of the World	— Arnold Kludas.
Ships of the Blue Funnel Line	— H. M. Le Fleming.
Blue Funnel Line	— Duncan Haws.
Blue Funnel Line, 1865–1914	— Francis E. Hyde.
A Merchant Fleet in War	— Capt. S. W. Roskill, R.N.
Marine News	— Journal of the World Ship Society.
Blue Funnels in the Mersey in the 1920's	— Sea Breezes Publication.
Sea Breezes. March 1950, April 1955, April 1960	
Blue Funnel and Glen Line Bulletins.	
Elder Dempster Fleet History 1852–1985	— Cowden and Duffy.

EXPLANATORY NOTES

The year given after each ship's name is the year the ship entered the fleet and is not necessarily the completion date; this date is given in the text together with the year of leaving by sale or transfer.

The names of the ships were all taken from Greek mythology, the same name being used over the course of time for several different ships.

DIMENSIONS, LENGTH x BEAM.
   Length. Up to and including 'Deucalion' 1970 the REGISTERED LENGTH is quoted.   This being some feet less than the LENGTH OVERALL (a.o.) which applies from 'Stentor' 1946 onwards.
   BEAM. Is given in feet.

G.R.T.   GROSS REGISTERED TONNAGE is liable to alterations during the life of the ship.

I.H.P.   INDICATED HORSE POWER is that usually given for steam engines.

B.H.P.   BRAKE HORSE POWER is the term used for a diesel engine.

S.H.P.   is that used for steam turbines.

The service speed is given in knots. Blue Funnel ships were no different from other liner companies whose ships were capable of speeds considerably faster than the service speeds quoted.